SURVEILLANCE AND SURVEILLANCE DETECTION

A CIA INSIDER'S GUIDE

T0039989

JOHN G. KIRIAKOU

Skyhorse Publishing

Skyhorse Publishing books may be purchased in bulk at special discounts for sales promotion, corporate gifts, fund-raising, or educational purposes. Special editions can also be created to specifications. For details, contact the Special Sales Department, Skyhorse Publishing, 307 West 36th Street, 11th Floor, New York, NY 10018 or info@skyhorsepublishing.com.

Skyhorse® and Skyhorse Publishing® are registered trademarks of Skyhorse Publishing, Inc.®, a Delaware corporation.

Visit our website at www.skyhorsepublishing.com.

10 9 8 7 6 5 4 3

Library of Congress Cataloging-in-Publication Data is available on file.

ISBN: 978-1-5107-5610-6
eBook ISBN: 978-1-5107-5615-1

Cover design by Kai Texel

Printed in the United States of America

For Charlie

Author's Note

All statements of fact, opinion, or analysis expressed are those of the author and do not necessarily reflect the position or views of the Central Intelligence Agency or any other US Government agency. Nothing in the contents should be construed as asserting or implying US Government authentication of information or Agency endorsement of the author's views. This material has been reviewed by the CIA to prevent the disclosure of classified information.

A Note on Redactions

Several passages in this book are blacked out, or "redacted." This was done by the CIA's Publications Review Board, with input from the National Security Agency's review board. They argue that these passages are classified and are not releasable to the public. Although we disagree, we elected to keep the redactions in the text, and blacked out, for the purpose of continuity.

Contents

Introduction

THE JOB OF EVERY CIA OPERATIONS officer is to recruit spies to steal secrets. It's really as simple as that. Every ops officer gets promoted, or fails, based on his ability to do this one thing. It's harder than it sounds. My CIA instructors told me on my very first day of ops training that I had to convince a target that I was his best friend to the point that he was willing to commit espionage for me. In some cases, he was committing treason for me. The stakes were that high. If we got caught, I would likely be expelled from whatever country I happened to be in. But my agent, the guy I had recruited, could spend the rest of his life in prison. In some parts of the world he could be executed. So I had to do my part to make sure that I was never, ever followed to or from a meeting. And I had to train the source, what the CIA calls the "agent," to do the same.

Surveillance and surveillance detection are an art form. After learning how to recruit spies to steal secrets, CIA operations officers learn surveillance detection because they are the two most important skills a person can have. Your life, safety, and security—and that of your agent—rely on your ability to safely get to and from a meeting.

The point of surveillance detection is to determine whether or not you are being followed by employing an increasingly sophisticated route, coupled with "normal" stops that a person would make in the course of running errands. Why? Because you want your surveillants to conclude, "This guy isn't doing anything unusual. He stopped to pick up his dry cleaning, he stopped to buy a bottle of wine, and he stopped by buy a shower curtain. We won't waste time on him anymore. We'll move on to the next person."

That's in the world of intelligence. But if you're an average citizen going about your normal business, maybe you want to know if a potential mugger is following you. Perhaps you want to know if your spouse has hired a private investigator to follow you. Perhaps the cops, or an insurance investigator, are checking you out for some reason. Whatever it is, you have the right to know about your surroundings and about who may mean you harm. That's what this book will teach you to do.

As I said, this is a skill that can be learned. You have to be calm, non-alerting, as we used to say at the CIA, and not raise suspicion. Fictional jewel thief A. J. Raffles explained to his less-experienced co-conspirator about surveillance: "But I *felt* him following me when I made tracks; though, of course, I didn't turn around to see." "Why not?" says the co-conspirator. "My dear Bunny," says Raffles, "It's the very worst thing you can do. As long as you look unsuspecting, they keep their distance, and so long as they keep their distance you stand a chance. Once you show that you know you're being followed, it's fight or flight for all you're worth. I never even looked around. And mind you never do the same."

With that said, you must always be aware of your surroundings. You must always pay attention. You must always be prepared for the worst.

Toward the end of my CIA career, I was assigned to a dangerous Third World country. Terrorist groups controlled the countryside and had a significant presence in many of the country's largest cities. Foreigners were killed there with some regularity. One day I left my small hotel to drive to the office. Just like every other day, I left at a different time and took a different route so as not to establish a pattern. I was about a quarter of a mile away when I noticed a guy on a motorcycle, wearing a red helmet, and trying very hard to stay in my blind spot. What bothered me the most was the helmet. I don't know where anybody would even *buy* a helmet in this country, and certainly, he was the only person I ever saw wearing one. He broke away from me about a quarter mile from my office. I was worried, but not panicked. Was it a coincidence that a helmeted motorcyclist had followed me to work? I wasn't sure yet.

I worked a normal day, which in the aftermath of the 9/11 attacks was fourteen hours. It was dark when I left the office. I only got about a quarter mile when I saw him again. He was wearing the red helmet and he was parked under a tree, waiting for me. I was frightened, but I still thought that it could be a strange coincidence. So I took a completely illogical route back to the hotel. I made several right-hand turns, forcing me to make several lefts. I overshot the neighborhood the hotel was in, forcing me to double back. And all that time, the motorcycle was on me. He broke off and drove away as I approached the hotel.

This was the very definition of surveillance: "Multiple sightings at time and distance." I had seen the motorcyclist more than once, at different times of the day, and at different places. I was under surveillance. The question now was whether the surveillance was hostile or friendly, such as the local police or intelligence service just checking me out.

The next morning, I woke up at 5:00 a.m., well before sunrise. I poked my head out the front door of the hotel and looked up and down the street, seeing nothing. I checked under my car for bombs and GPS devices, again seeing nothing. I got in the car by 5:30, pulled out of the gated hotel property, and made a right, heading into a residential neighborhood, rather than a left toward the highway that would have taken me directly to the office. And there he was again.

I worked hard to remain calm. Trying to lose him would have alerted him to the fact that I knew he was following me. I had to make him feel comfortable. I had to make him think that I believed it was normal for a helmeted motorcyclist to be following me to work for days at a time and at 5:30 in the morning. I took an odd route to the office again and, again, he broke off about a quarter of a mile before my destination.

I was practically panicked. I had to assume that the surveillance was hostile, that the driver meant me harm. I waited an hour for the security officer to get to work. When he did, I went straight to his office. "I'm under surveillance," I said. "I'm positive." I recounted to him the three sightings, the locations, and the times. He agreed that I had a problem and said we needed to talk to the chief right away.

The chief arrived an hour later and, trying to control my panic, I repeated the story to him. He listened with concern and finally said, "Well, you know what you have to do." I did. I would sign out a gun from Security. "If I see him again, I'm going to kill him," was my response.

That afternoon, I had a meeting with members of the host country's intelligence service. It was a routine meeting of paper exchanges and information sharing. On the way out, I had a thought. I turned and said, "General, let me ask you a question. Are you following me?" He said no and asked why. I told him, "I'm under surveillance. I'm *positive* that I'm under surveillance. And if it's not you or your men following me, I have to assume that it's hostile. If I see him again, I'm going to kill him. There won't be any warnings."

I never saw him again. I had to assume that my foreign friends were curious to see how I was spending my time when I wasn't liaising with them. And they chose somebody to follow me who was terrible at surveillance. I actually saved the guy's life by asking that off-handed question. As frightening as the experience was, it made me realize that I was good at surveillance detection. Very good.

I became a surveillance detection instructor when I returned to CIA Headquarters at the end of my tour, teaching young CIA officers yet to be stationed overseas how to stay safe and alive by spotting surveillance. And when I left the CIA, I created and taught surveillance and surveillance detection classes at the university level. As I said earlier, there are real-life applications for this kind of skill outside of the CIA.

This book will walk you through the steps of surveillance and surveillance detection, including how to craft a sophisticated surveillance detection route, how to identify appropriate surveillance detection route (SDR) stops, and how to conduct yourself in a non-alerting fashion. It will teach you to be secure in the knowledge that you are not (or are) being followed. It'll teach you how to construct a professional SDR and how to conduct surveillance on your own.

We will use maps and Google Earth images to plan the perfect route in advance. And we'll mix things up with alternate modes of transportation. Remember, the goal is to get from Point A to Point B safely and securely, being so boring and routine while you're doing it that the cops/bad guys/private investigators/jealous spouses and others lose interest and move on.

CHAPTER I

What Are Surveillance and Surveillance Detection?

THE DEFINITION OF SURVEILLANCE IS EASY. It's multiple sightings at time and distance. That means that you have seen a person or people following you, or watching you, more than once, at different times, and at different places. It could be on the way to work and on the way home from work. It could be while you're out shopping or going to the gym. It could be in church or at the mall. Surveillance detection is simply the act of spotting that surveillance. You do that by employing an increasingly sophisticated route, coupled with "normal" stops that a person would make in the course of running errands.

So what are the surveillants looking for? Why would they be interested in you? The first thing a professional surveillant will look for is demeanor. Do you look suspicious? Are you looking over your shoulder to see if you're being followed? What does your body language look like? Are you walking more quickly than anybody else? Are you speeding or driving aggressively? Are you changing clothes in the middle of an SDR? Believe it or not, this can sometimes be a good idea, particularly if you're trying to *lose* surveillance. But under normal circumstances, nobody goes into a store wearing a blue shirt and jeans and comes out wearing a red shirt and khakis. Destination also matters. In the intelligence world, a surveillant is interested to know if you're going to a hotel in the middle of the day or to an obscure restaurant far off the beaten path at an odd time.

Your goal is to act as normally as possible, at least in the intelligence world. The theory is that if you appear to be a normal person, going about your normal business, the surveillants will move on to the next person and leave you alone. Is it really worth their time to spend a day following you to the dry cleaner's, to a store, to buy groceries, when they could be spending time trying to identify a spy? No way.

I once worked overseas with a woman who was a State Department political officer. Her job was to do normal diplomatic stuff with the host government—"carrying out the diplomatic business of the United States," they call it. But the host government just automatically assumed that she was a spy. Why? I have no idea. She tended to be very arrogant and that probably rubbed them the wrong way. One evening she was on her way to a diplomatic cocktail party, and she noticed that she was being followed. She thought she'd have fun with the surveillance team, which was almost certainly from the host government's intelligence service. So she sped through the capital city, driving the wrong way down one-way streets, driving through residential neighborhoods like a bat out of hell, and going all the way outside the city limits and then doubling back. In the end, she arrived at the cocktail party, smiled at the surveillants, and went inside. Cute, right? She certainly thought so. But what she did is convince the local intelligence service that she was a CIA officer. They were on her like white on rice for the next two years. They made her life miserable. Everything she did that night was wrong, wrong, wrong, even if in the end, she diverted the intelligence service's attention away from CIA officers and onto herself.

In the intelligence world, a surveillance detection route (SDR) is done in advance of three operational acts: 1) An operational meeting, where an operations officer is going to meet with a source. The meeting time and place have been predetermined, either because it was triggered based on protocols or it was regularly scheduled; 2) A brief encounter, where an operations officer quickly passes or receives a message from an agent in a meeting that takes literally seconds; and 3) A dead drop. This is the most difficult and dangerous operational act. There are several phases to a dead drop. First, either you or the source has to signal the meeting. Second, you have to drop something (a message, a thumb drive, or a cell phone, for example). And third, you have to signal that the drop is completed. Every one of these three phases requires an SDR. There is more planning involved with a dead drop than with any other operational act.

Surveillance is not necessarily a couple of guys in a car, following you. It can be a lot of different things. In the intelligence world, surveillance can be a whole bunch of people in a whole bunch of cars following you, dropped off and then picking you back up later. All the while, they're in touch via walkie-talkie. But surveillance can also be static, that is, stationary. It can be closed circuit television, webcams, or even an apartment building doorman who just stands there and watches cars go by.

There are a lot of different kinds of surveillance and surveillance detection. It can be via car, truck, or motorcycle. It can even be by plane. It is often on foot. Perhaps the best kind of surveillance detection route is multi-modal, especially if you have an operational act that you absolutely have to accomplish and if you don't care if the bad guys know who you are. CIA officers overseas will often use multi-modal surveillance detection routes. You might start out on foot, make a stop or two, and then take a bus or subway across town. There you might get into a taxi, or even a rental car, and use that to complete your route. As you can see, it's provocative. But if you really want to lose surveillance, having them all following you on foot when you then get into a car and drive away may be the way to go.

Not all surveillance has to be this sophisticated. Authorities everywhere like things called "red zones." These are areas where the security and police presence are so prevalent that they don't even need to follow you. There are cameras everywhere. Cops or private security guards are everywhere. Think the White House, the Pentagon, or CIA Headquarters. If you were engaged in something where you did not want to be detected, you would be a fool to go anywhere near a red zone.

Surveillance can also be non-traditional. A foreign intelligence officer once told me a fascinating story. He said that his service had been tasked with surveilling a nuclear scientist from an enemy country who had arrived to attend an academic conference. The scientist, of course, had been fully briefed by his own country on hostile surveillance. When he arrived, he was practically paranoid, doing a sophisticated surveillance detection route and constantly looking over his shoulder—behavior that was certainly "alerting." But the local intelligence service was ready for him. There was nobody following him. Literally nobody. Instead, the service had put seventy-five people on the case. But they were all walking *toward* him. After each surveillant passed, he or she simply doubled back and walked toward the scientist again. Because he was looking behind him, he didn't notice that all the people walking toward him repeated themselves. In the end, seventy-five people accompanied him to his operational act. The scientist was arrested and expelled. And his local handler from the enemy embassy was identified and expelled.

That's what successful surveillance detection looks like from the side of the surveillant. Your goal, though, is to make sure that you don't attract attention to yourself, that you don't get caught on your way to whatever it is that you are trying to accomplish, and that you are able to go about your business unmolested. This is going to require a lot of planning. And by the end of your SDR, you will be certain that you are either clean of surveillance or you're being followed. The choice, then, is yours. You can abort your mission, continue on to your meeting, or try like hell to lose the tail. Ninety percent of the work is in the planning.

CHAPTER 2

Creating a Surveillance Detection Route

SURVEILLANCE DETECTION ROUTES TODAY ARE FAR easier to come up with than they were a decade ago. That's because of the advent of Google Earth, Google Maps, and GPS. When I first went through surveillance and surveillance detection training at the CIA, there were no such things. We planned out our routes with paper maps that we then cut and taped together to create the entire route. Now the planning is much easier.

There are four basic phases in a surveillance detection route. The first is the *kick-off phase*. This is where you begin the SDR, and it is usually your home or office. Everybody leaves home or work, right? Usually every day. There's nothing at all suspicious about it. You leave home and, taking a normal road, begin to go about your normal business. You begin by taking a "red road." That's a major road, one that appears on a map in the color red. It's a road that everybody going about normal business would take to get from home or work to point A. It's virtually impossible to tell at kick-off whether you're under surveillance. Sometimes you get lucky, but that's not really your goal here. Your goal is just to head out, going about your business.

For phase one, you will travel from your kick-off location to your first stop. That stop can be anything that *makes sense*. It would make sense to go to your local bank, the nearest grocery store, or the dry cleaners. It would not make sense for you to drive fifteen miles to a 7-11, when there are twenty other 7-11s between your kick-off location and your first stop. It would not make sense for you to go ten miles to a grocery store when there are a half dozen other grocery stores closer to you. It would, however, make sense for you to travel a distance to go to a specialty grocery store. Maybe there's an Asian place that has a certain kind of fish you can't get anywhere else. Maybe there's a Brazilian place that specializes in unusual cuts of beef. But for the most part, it's better to keep it simple.

Phase two is the *pattern phase*. If it makes sense, you can establish a pattern to your stops in this phase. Maybe you're going to a series of antique shops today. Or coin dealers. Or auto parts stores, looking for an unusual part. It is in this phase that you will begin using "black roads," that is, roads that appear in black ink on maps. They are less used. They cut through neighborhoods. Black roads make it far easier to detect surveillance. So when you come out of your first stop, the one on the red road, you then turn into a nearby neighborhood to cut through on your way to your next stop.

Pay attention when you turn off. Did anybody turn off with you? Was there anybody waiting for you to come out of the stop? Did anybody go into the stop with you to see what you were doing? I once used a specialty wine shop as my first stop. I was in there for about fifteen minutes, legitimately looking for a vintage port. Apparently, my surveillants thought I was up to something. So they sent a young woman in after me. She knew nothing about wine. The shop wasn't very big, so everybody noticed that she had entered. Finally, the owner asked if he could help her. Because she knew nothing about wine, she looked at him blankly and asked where he kept the Manischewitz, a Jewish dessert wine drunk during the Passover seder, which had taken place six months earlier. She made a fool of herself. This was a high-end wine shop, not the Walmart seasonal aisle. Not only did the puzzled owner send her on her way, she raised her profile with me. I knew immediately—only twenty minutes into the SDR—that I was under surveillance, and I aborted my meeting. I paid for my wine and drove home. Pay attention to your surroundings, but also use your judgment to determine the appropriateness of your stops.

Phase three is the *aggressive phase*. This is the SDR phase where you become increasingly aggressive in your route. That's not to say that you start speeding or otherwise drawing attention to yourself. It means that you begin making more obvious moves to determine your status, to see if you're being followed. You will use only black roads in this phase.

One of the techniques you will employ in this phase is called "stair-stepping." That means that instead of driving or walking straight for ten blocks and then making a right, you drive straight for one block, make a right and go one block, then make a left and go one block, followed by a right and one block. Your pattern looks like a staircase. And every time you make a right, and you look both ways to see if you can safely turn, you are actually looking up and down the street for active surveillance.

It is also in this phase that you might use cut-throughs in residential neighborhoods. You might drive through a strip mall parking lot or use a cemetery as a cover stop. Buy flowers at an earlier stop and lay them at somebody's grave. Go to an obscure, small museum and take a tour. If you have time, go to a movie. Better yet, go to a movie and leave forty-five minutes into it. Go to a party store and buy a bunch of balloons. You might do something that makes a surveillant scratch his head, but is explainable if you're asked about it.

Phase four is the *provocative phase*. By the time you enter this phase, you should be 99 percent certain that you are not under surveillance. This phase doesn't make any sense to a surveillant at all. You might drive or walk half-way down a street, turn around, and drive or walk back. You may pull into a driveway, turn around, and then do the same thing again 100 feet down the street. Your goal is not to convince someone who's following you that you're headed to a store. It is simply to make one final determination that you're free of surveillance. Once you are 100 percent certain that you are free of surveillance, go on to your final stop, which should be in the same area.

CHAPTER 3

Surveillance Concepts

THE CONCEPT OF SURVEILLANCE IS ONE of the oldest of all civilizations. Numerous Biblical accounts tell of the use of surveillance, sometimes to help the community and at other times for the benefit of an individual. In 2nd Samuel, there is the story of David watching Bathsheba, the wife of Uriah, and planning on how to steal her from her husband. By contrast, Numbers 13: 17–20, provides guidance on the work to be performed in surveillance. They were instructed to collect information and the terrain, the local population, the quality of the soil, and whether or not it supports trees. This is the vital information that would enable leaders to determine the vulnerabilities of an adversary and how to exploit those weaknesses. This is consistent with the instructions given to modern spies and supports assessments needed for decision making.

As was seen in the book of Numbers, spies are supposed to collect information. However, history shows that they can also spread false information to mislead their enemies. In his book *The Enemy Within: A History of Espionage*, Terry Crowdy writes that the earliest surviving record of surveillance and espionage dates back to 1274 BC when the Pharaoh Rameses was at war with the Hittites and sent operatives pretending to be deserters from the Pharaoh's army. The Hittite king sent two spies to the Egyptians with instructions to convince the pharaoh that their army was too far away to pose a threat. Rameses was convinced and sent a small contingent of his forces into the field. As they encountered a large Hittite army much closer to their position than they expected, Rameses' force was overwhelmed.

Histories of the American Civil War focus on the role of cavalry surveillance units that were supposed to observe but not engage enemy forces. This is consistent with the traditional view that surveillance is an information collection tool. The starting point is the clarification of key points in an operation. The clandestine activity, including a surveillance mission, will be authorized by a specific organization which will be designated as the sponsor. This might be broadly stated as a government itself or, being more specific, a particular agency of the government. Most of the activities considered in this book are conducted by the Central Intelligence Agency.

When dispatched by a governmental agency, the surveillance operation will concentrate on a specific nation, region, or a person. These entities against which clandestine activities are directed will be identified as the targets.

An entity will be designated as a target if it is seen as a threat to national security. If an organization, group, or installation is seen as a threat to national security, it will be designated as the operational target. This goes for an important foreign national who could undermine our national security. He would be regarded as a threat target. An individual who is has great influence over an operational target is designated as an "actively involved personality." Influence can be determined from the individual's rank or his participation in fund-raising or recruitment. Should an individual's participation be limited to open, legal mass activities rather than underground activities, he will be viewed as a "nominally involved personality."

When the sponsor indicates the requirements for this activity, a mission analysis must be conducted. The first step is to determine if the team has the capabilities required for the mission. Next, the collection tasks must be identified in terms of those which are explicit as well as those that are implied. If the operation is to take place in a distant region, relevant transportation capability is an implied task, for example. A third step is to conduct a search through existing data bases to be sure that these tasks have not already been accomplished at another time by another team. If, after an examination of the intelligence collection requirements, it is clear that that information has not been collected previously, then it will be time to perform a target analysis. This will involve a categorization of the location of the objective and the difficulties associated with penetrating the facility that contains necessary information.

There are several important aspects of surveillance that must be explained. As we'll discuss, a distinction must be made between the concepts of close surveillance and loose surveillance. The former refers to continuous observations of a target. Even if the surveillants are detected by their target, the observation is not stopped. By contrast, loose surveillance is not continuous and may be stopped if the subject becomes aware of the surveillance team.

If the target is not expected to move, the most useful technique is fixed surveillance. This option is one that does not require more than one surveillant. Ground surveillance is utilized when there might be a particular route that needs to be monitored. For example, if there is a bridge that might be used as a dead drop for enemy agents, ground surveillance would focus on that particular location while waiting for something to happen.

Another concept is that of foot surveillance. This type of surveillance is not likely to be detected unless the surveillant wants to be seen or is just poorly trained. One category of foot surveillance, generally used when the target is stationary, is one-man surveillance. While watching in this setting, the surveillant not only observes a target but he also endeavors to collect information related to the target or his place as part of a larger operation. If two men are employed when the target is moving, one surveillant should be behind the target while the second surveillant observes from across the street or at a distance. The best option in foot surveillance is utilization of a three-man team. Surveillants one and two are behind the target while surveillant three in across the street and slightly to the rear of the target. If the target makes a turn onto another street, surveillance one continues in the same direction while surveillant two will turn to follow the target.

Surveillant three will then replace surveillant two and resume the observation.

A simple but effective variation on this procedure is known as leap-frog surveillance. This is useful for preventing detection by the target since the surveillants rotate their positions so the target is unlikely to recognize a surveillant because he sees him too often.

If the target is someone who has a regular routine, leaves home at the same time to go to his office, and returns home at the same time each day, progressive surveillance is an appropriate technique. This is something that can be done on a daily basis for as long as the target sticks to a routine.

Vehicular surveillance can normally use anywhere from one to four vehicles. If only one vehicle is available for this operation, the surveillant will follow the car in which is target is traveling and must determine the safest distance to assume behind the target vehicle. If the target is moving in an urban area in which streets are straight and parallel, two-vehicle surveillance can be employed. Vehicle one is behind the target vehicle while surveillance vehicle two is on a street that is parallel to the one used by the target and surveillance vehicle one. The most effective technique for urban areas is four-vehicle surveillance in which vehicle one is in front of the target and vehicle two is behind him. Vehicle three is on a parallel street to the right while vehicle four travels on a parallel street to the left. The target is effectively boxed in by this technique.

If the objective is broader and focused on an area, the appropriate type of observation is known as reconnaissance. This is a form of surveillance but is different from the traditional notion of surveillance. The goal of a reconnaissance team is activities and resources of an adversary while collecting information about a specific area. This information would include geographic aspects of the area and anything that would affect future operations.

When future clandestine operations are being considered, the reconnaissance process will be described as casing. This is a relatively casual observation of an area that might be selected for an operation. When the Americans and British were considering the construction of a tunnel into the Soviet sector of Berlin, they had to study the neighborhood in which the tunnel would start. Their casing of this area meant observations about the sort of people in that remote part of the Western sector of Berlin, their activities and behavior, the nature of the soil, and the extent to which the location could be seen by East German sentries. Prior to starting the project in Berlin, the CIA practiced by digging a similar tunnel in ██████████ where the soil was similar to that of Berlin. Three decades later, when the CIA was studying how to gain access to cables linking the Soviet Ministry of Defense with a research institute in a closed Moscow suburb, CIA operatives spent two years casing the area and photographing a likely penetration point. In performing work such as this, maps are useful but the best technique is to send an individual who will be working on the operation.

Photography is a vital part of most surveillance operations. When the American, British, and French military liaison teams were working in the Soviet zone of Germany in the 1950s, their leadership would not accept mission reports that were not accompanied by relevant photographs. This was a time when drawings—once viewed as proof of a siting—were being phased out of reports because of the increasing availability of effective cameras that could be used in surveillance operations.

For intelligence operations, there are several types of photographs that are needed. Technical photographs are especially important because a verbal account of a complex instrument would be of limited value. Getting photographs of technical plans is an important objective for any intelligence operation. There is also a great need for photographs of individuals that are high value targets for an intelligence or military organization. In the second Gulf War, photographs of individuals sought by coalition military authorities were placed on playing cards distributed to all military personnel. A clear photograph should show the head and highlight key facial characteristics. If the photo is posed, the camera should be six feet away from the subject. While post-World War II photographic technology was less advanced, "wanted posters" were prominently displayed so people could recognize the war criminals who were objects of Allied searches.

In a broad sense, effective photographic methods are needed to provide documentation about actions taking place in the field. Special methods, such as long-range photography, have been developed to enhance this aspect of intelligence work. Innovations that improve photography in a dark environment have contributed to these operations. Regular photography has been supplemented by video technology while digital photography has led to the creation of cameras with an enhanced storage capacity.

A goal of surveillance is usually the visual acquisition of information. However, information can also be acquired by simply listening to what people say. With this, the operative can gain information through subterfuge, so the target has no idea he is providing useful information to a surveillance team member. This is an indirect process that begins with a casual conversation in which the surveillant establishes rapport. The setting for such an endeavor is important. If the operative is able to encounter a subject in a bar, that represents a good environment. Buying drinks for the target is a good way to establish rapport and get him to speak more freely. The reduction of inhibitions can facilitate indiscreet talk in such a setting.

Sports, weather, or jokes might be the starting point for undertaking this kind of verbal surveillance. At this stage, the operative might act as if he is an expert on all things, somewhat like the bombastic character Cliff Clavin on the popular comedy *Cheers*. His know-it-all manner could prompt the target to counter with accurate information simply to put the know-it-all person in his place.

Another possible approach is to flatter the target by implying that his opinion on all matters is important. While you begin with a non-relevant subject you can eventually draw the conversation into an area that is consistent with your surveillance interests. Such flattery may encourage the target to speak more freely since he sees you as an admirer. A contrary strategy is to adopt the position of a cynic who disputes everything he hears. The surveillant's hope is that this might prompt his target to elaborate on his observations as a way of defending his ideas.

You might also spark the subject's enthusiasm by appearing to lose interest in what he has to say. Sometimes, if the operative seems disinterested, the subject may respond with more information. Or, if the operative remarks that nobody is able to understand that particular topic, the subject might try to demonstrate that he does understand.

One key to successful verbal surveillance is to identify the correct verbal device or phrases that will direct the conversation. By listening carefully to a potential source, you can identify the probes that might be effective. A general question to clarify a point the target might have made could be useful. You indicate your agreement to whatever point he made but ask him to clarify some aspect of his statement. If your source makes a statement that is vague or incomplete, you might indicate that as a non-specialist the concept is beyond your intellectual grasp. Therefore, you can ask for another explanation to help you.

If none of these techniques have worked for an operative, he may decide to become more confrontational. He might point out that there is a contradiction in which the subject said and ask him to resolve this matter. Of course, there is a danger that instead of resolving the contradiction, he might simply contact the security services. Therefore, this technique is only used as a last resort and when the operative is ready to flee if confronted.

These are narrow categories that reflect specific objectives of surveillance. There are, however, broader and more general categories that are more reflective of the technologies being used. Most discussions of surveillance will begin with recognition of the ubiquitous Closed-Circuit Television (CCTV) systems. The CCTV is a necessary component of visual surveillance. It is a fairly basic innovation that has revolutionized large-scale surveillance procedures. Not requiring the services of surveillance teams, this is a cheap way to monitor large areas and have a record of what was seen. Visual surveillance is used to detect and track individuals or objects. They may be in movement or stationary. The CCTV creates a new dimension to surveillance because it is continuous and omnipresent. It does not require a specific surveillance order, but will often provide useful information after the fact when there has been an event or activity of interest to authorities. It is, in short, as relevant to criminal investigations as it is to intelligence operations. Yet, it is a reminder of a negative aspect of our new technologies because it produces such an abundance of data that it is increasingly difficult to process. Having too little information about a target is a difficult situation. Having too much information may be even more difficult.

Another broad type of surveillance involves the use of biometrics. Biometric identifiers are used to note those physiological characteristics of an individual that make him unique. Some of the more common characteristics are fingerprints, facial recognition such as the shape of ears, DNA, palm prints, and retinal recognition. Many airports today will require a retinal scan for recognition of each individual. More colorfully, during interrogations, East Germany's Stasi, the country's main intelligence service, would have suspects sit on a cloth stretched across the bottom of a chair. The cloth was stored in a sealed jar upon which was placed a label with the subject's name. Specially trained dogs would help identify people on the basis of their recorded smell.

One of the most reliable of biometric technologies is the fingerprint. A fingerprint is an impression left by the ridges of a human finger and the individuality of this data is such that there is a statistical improbability of any two fingerprints being identical. We might think of this technology as being recent, but during the time of the Babylonian king Hammurabi (1750 BC) authorities would take fingerprints of people whom they arrested. In the United States, it was not until 1902 when fingerprints were used for the identification, arrest, and conviction of a murderer that fingerprinting came into common use. Over the years, fingerprints have gained acceptance as one of the best ways for identification and the UK government has long debated fingerprinting all children aged eleven to fifteen and creating a data base using the fingerprints.

Retinal recognition has increasingly gained acceptance as a foolproof method of identification. Formation of the retina takes place before birth, and it remains the same throughout a person's life. It has over 266 distinctive characteristics. Retinal recognition is a relatively new method of confirming the identity of a person by analyzing its random patterns.

There are several shortcomings associated with biometrics. One of the most frequently cited is the issue of how such data is acquired, since much of it can be compiled without a subject's consent or knowledge. Biometric indicators are immutable, so once a person's information has been compromised, there is no way of compensating for this. Equally troubling is the possibility that such data might be utilized for identity theft. Recent data breaches have shown how easily personal data can fall into the hands of criminals.

"Dataveillance" is the final type of surveillance in these broad categories of technical surveillance. The demands of modern society have forced people to yield more and more personal information to computerized systems that store a person's entire life story in data banks that may or may not be secure. During the COVID-19 era, with its lockdowns and quarantines, even more of our transactions have been directed through advanced technologies. If you note the ease with which you can make purchases through DoorDash and other conveniences, you should realize how much is known about your interests, appetites, and payment methods. Major retailers have long created "loyalty cards" that enable them to anticipate your future purchases and understand every aspect of your economic situation. Airlines already provide personal information about travelers on their flights so the government can determine if any of them represent a "terror threat." This raises the possibility that an authoritarian regime might use the law to require companies to obtain information about their clients if the government deems such information essential for "national security."

CHAPTER 4

Understanding Hostile Surveillance

A S YOU CREATE A SURVEILLANCE DETECTION route, it is important to note other vital issues. One of the most consistent concerns is to identify hostile surveillance. Ami Toben, who is a pioneer in the field of covert surveillance, defines hostile surveillance as the covert observation of a target for the purpose of collecting information.

Hostile surveillance is a covert effort to look at a target and determine its value. It will also study security procedures that are being used to protect the target. Finally, the hostile surveillance team will identify vulnerabilities and escape routes that can be used after the hostile action has taken place.

There are several components to this process. The first is that it involves physical observation that is done covertly. During the Cold War, the United States Military Liaison Mission, along with its British and French counterparts, patrolled East Germany but were, under relevant postwar agreements, required to do this overtly. That was a crucial point because it meant their activities were officially classified as mere observation rather than espionage. Their tours through East Germany were not classified as surveillance. By contrast, physical surveillance must combine observation with covertness if it is to be regarded as proper surveillance. The goal of hostile surveillance is to collect information that can be used in planning an attack against a target.

There is generally some type of preoperational hostile surveillance that is conducted prior to an attack. Preparing for such an attack is a long process that begins with a member of a surveillance team acting as a scout moving through the neighborhood in which the target is located. This person should study the general environment in order to advise his team on how to blend in with the people seen near the target. If it is located next to a university, it is likely that the people on the street will look like students. If members of the team dress like homeless people, they will be conspicuous and be spotted by the most basic security forces protecting the target.

The scout should identify vantage points that can be used for long term observation of the target site. The vantage point should be one that gives the operative a broad view of the target itself so he can collect visual information about it such as the types of security measures that are in place. Before selecting a vantage point, the scout should find a location from which the vantage point can be studied. This location should not be one from which the target is seen since the goal here it to learn about the vantage point. If it is in a park, he needs to know what sort of people are normally in the park so his team members can assume the correct identities. If the vantage point is a bench, he must determine what sorts of people normally occupy the bench during proposed observation times.

A well-publicized example of an incident that highlighted the significance of hostile surveillance and the need for counter surveillance efforts was the so-called "mansion murders case" in 2015. It began with a fire in an exclusive home in the Woodley Park neighborhood of Washington, DC. As firefighters fought to suppress the blaze, they found the bodies of three adults and one child, all of whom had been murdered. The home belonged to Savvas Savopoulos, the CEO of a construction company known as American Iron Works. The ten-year old Savopoulos child had apparently been tortured in an effort to get his father to comply with demands made by the intruder. What the killer sought was apparently $40,000 in cash, since that amount of money was later delivered to the home by a Savopoulos assistant. Shortly after the funds arrived, the home was set on fire and a sports car owned by Savopoulos was found burning in a church parking lot in Maryland.

While Savopoulos and his family waited for the money to be delivered, which took until the next day, two pizzas were delivered to their house. The pizza was instrumental in the apprehension of Daron Wint, whose DNA was found on a partially uneaten pizza crust. Wint was a former employee of American Iron Works and had been fired for threatening another worker with a knife in 2005.

There was a decade between the time Wint was fired and his attack on the Savopoulos home. As Wint had difficulty maintaining a job, he apparently focused on his being fired by Savopoulos as a key event in his troubles. A home invasion such as this required considerable planning and was not the result of an angry impulse. Investigators noted that in 2010, an armed Wint had been arrested at a gas station across the street from the American Iron Works. Such a location could serve as a "surveillance perch" for Wint if he intended to follow Savopoulos home after work.

A home invasion such as this requires extensive surveillance in order to determine patterns associated with the home and the neighborhood. An attacker would need to learn what protective measures existed and how to best enter the home and control the occupants. This is all part of the attack cycle and it is during this time that a would-be attacker is vulnerable to observation. Had Savopoulos developed any plans for how to protect his home, he or a security employee might have observed someone watching. Even a personal situational awareness effort might have saved him and his family. Because Wint had no professional surveillance training, he would likely have made obvious mistakes, but since no one was looking, he remained unseen.

In a city such as Washington, there are numerous companies that provide counter surveillance services, and if he had employed one, the attack could have been prevented. Many executives routinely employ counter surveillance personnel to protect themselves. One simple intelligence tool that would have been useful is a *duress code*, a simple warning word that could have been included in his communications when Savopoulos communicated with his staff to arrange for the collection and delivery of the money for Wint.

There is no doubt that Wint, as a non-professional, likely used poor techniques and could have been easily detected. But that could have happened only if someone had been observing and practicing situational awareness. There are four major elements in training people to be observant in protecting their interests or their clients. They are time, environment, distance, and demeanor. Together, trainers use the acronym TEDD as a reminder of these vital concerns. The starting point for this equation is to notice whether you keep seeing the same person or people over a fairly long period of time. If someone "just happens" to appear as you travel to work each morning, perhaps in the same subway car, and later as you are having lunch, you need to remember this person. It is a clue that you may be under surveillance. If an individual thinks he is being targeted, he should then consider the other TEDD factors. TEDD is relevant for an individual who is being targeted for a complex home invasion such as in the "mansion murders." None of this applies to a situation in which a location, rather than an individual, is the target. A person who had the bad luck to visit the World Trade Center on September 11, 2001 could have done nothing to protect himself. On the other hand, a person who happened to be at a concert that was hit by an armed terrorist might have been able to note the suspicious demeanor of the terrorist planning to attack the location.

Of course, a trained terrorist conducting surveillance will be better able to avoid detection by changing his appearance. A change of clothing, even a reversal of the jacket, or the application of a wig, will do a lot to make the surveillance team member harder to see. Because of this, a counter surveillance person should pay more attention to the demeanor of individuals in their area. Even more important is to note the person's build, facial features like ears or noses, and how a person walks. These are things that cannot be easily changed.

There will always be critical places that should be watched in order to detect hostile surveillance. The gas station across the street from the American Iron Works was such a location. It gave Wint a good view of the factory and allowed him to see the choke point created at the exit from the factory. This was a critical place for someone hoping to follow Savopoulos to his residence. The front door of the office or home of a target is another critical location for hostile surveillance efforts. By identifying such places, they can be monitored by teams protecting a possible target. Finding trash or an assortment of cigarette butts might be an indication that the perch was being used for hostile surveillance.

One way that a hostile surveillance team member might expose himself is through improper demeanor. A person who is spying on a possible target must fit in the environment even though he knows he is an alien presence. This uneasiness can be hidden by a trained professional, but usually not by an amateur. Untrained surveillance operatives will react when they fear they have been seen. It is natural that they will try to hide their faces or abruptly turn around if they encounter their target. A well-trained person will be able to hide their normal reactions even when they are stricken with fear of exposure. An unexpected jump or an exclamation could insure their capture by security personnel.

Maintaining proper cover will contribute to the success of a surveillance operation. There are two aspects of cover. The first is cover for status, which refers to one's cover identity. This could be obvious identities such as businessman, student, or repairman. The second is cover for action which explains what a person is supposed to be doing. It could be as basic as sitting on the bench at a bus stop or attending a conference.

To have credibility, the components must be consistent so that the person conducting surveillance looks normal and fits into the environment. If the surveillant happens to be wearing a suit when his target goes to the beach, consistency is lost. A cover for action would fail if the surveillant appears to be walking his dog but has none of the gear needed to pick up after the dog, simply does not allow the dog to stop and do his business, or obviously is not liked by the dog.

Other demeanor failures would be getting up and moving immediately after your target moves. This creates a clear impression that you are following that person. Making a cell phone call each time the target moves is another example of a demeanor failure. Even more obvious is showing that you are part of a team by gesturing to other people in the area. Of course, there are times when the objective of a hostile surveillance team is to make sure the target knows he is being followed. The purpose of this could be, first, to harass the target. US embassies in Eastern Europe during the Cold War reported that hostile surveillance teams would push embassy personnel into the street. A second purpose of this type of surveillance could be to prevent their target from servicing a dead drop or preventing a meeting between a handler and his agent.

There are numerous demeanor indicators, many of which are almost imperceptible. If a person appears to fidget when you look at him, this might be an indication that something is wrong. It may not indicate criminal intentions, but it is still suspicious. A person's eyes can also be a demeanor indicator. A stubborn refusal to make eye contact or eye contact that conveys hostility are warning signs that someone might be a threat and should be avoided if possible.

CHAPTER 5

A Surveillance Detection Response

A SURVEILLANCE DETECTION RESPONSE CAN VARY FROM simple things done by the individual to a complex plan managed by professionals. But before moving toward formal, calculated responses, there are some rather simple responses that might be effective. They are obvious, common sense approaches.

Ami Toben, a professional who makes money developing and managing complex plans, suggests several low cost/no cost options. A first step is to demonstrate that you practice situational awareness and will not be a soft target. This means avoiding areas that seem to invite danger. It will always be dangerous to travel through a dark alley where pre-attack surveillance could begin. If you feel that someone is paying more attention to you than is dictated by your physical appearance, look directly at them. As you do this, it is likely that the observer will go away. If the surveillant does not look too dangerous, a personal confrontation might be sufficient to cause him to back off. If you tell him you will call the police, that will add to the credibility of your response.

Maintaining a low profile is another way of avoiding surveillance. Exhibiting flamboyant behavior and driving around in expensive automobiles will attract attention that is better avoided. If possible, it is a good idea to vary the routes you take going to work or any other place you routinely frequent. Sometimes this may not be possible. During a terrorism scare, personnel stationed at the Special Operations Command (Europe) located at Patch Barracks in Stuttgart, Germany were instructed to vary their routes when coming to work.

Unfortunately, only one road served the main entrance to the post, thus making this impossible. In order to mitigate that risk, soldiers were directed to go to work in civilian clothing and change into uniforms in the office. This still did not make much difference.

However, some modest variations will help. One example of this would be if you take care to enter your office building through a different door from the one you use leaving the office. Another helpful practice is to stay on the move as much as possible. This means you should avoid positioning yourself in a static location for an extended period of time. If you decide to have lunch in a sidewalk café, it is likely that you can be observed from a great many comfortable locations for surveillance personnel. If you arrange to meet someone, avoid meeting them at an outside venue. It is much safer to meet inside a restaurant at which you and a colleague may have lunch. Equally effective, if you have the time and are in an appropriate location, is to spend as much time as possible in one place. Surveillance is often extremely tedious and not the least adventurous. If he is not a highly trained, well-paid professional, you might be able to bore the surveillant into leaving.

A variation on this is to establish a comfortable pace in which it will be easy to keep track of you and will reassure the surveillant that you are an easy target. You want to be noticeable and seem predictable. Look for a site at which you can break his view of you for just a few seconds. This could be something as simple as turning a corner. The time when you cannot be seen will give you an opportunity to change your appearance. Change you head covering, your shirt, or your manner of walking. By the time the surveillant expects you to reappear, your physical indicators will be different, and you will be lost to the surveillance operative. In this respect you are like an illusionist who manages to make his audience look in the wrong direction.

Yet another way to end the surveillance is to simply move away quickly. This can be either at a run or a fast walk. It is important that you don't want to flee into a remote isolated area. If you go into a crowded store or get on a bus that is about to leave, you become a more elusive target. Perhaps the surveillant will decide you are more trouble than you are worth. This is another example of what is sometimes called *surveillance evasion.*

Keep in mind that by itself surveillance is not more than the collection of information. Although this will make most people uncomfortable, it is not a threat until it is linked to a hostile plan. It is reasonable to assume that if some entity is devoting the time to collect this information, there is a motive behind the activity. At this point, you need to move beyond the overt responses that you have made on your own.

After you have assessed your situation and determined that you are under surveillance, it is necessary to prepare your surveillance detection response. If Savvas Savopoulos, the victim in the 2015 "mansion murders," had developed such a response, he could have survived the home invasion that led to his death.

This plan should deal with five basic issues. The first is what type of SD will be employed? It can be static or mobile. The second issue is when will the SD be conducted? Will it run for twenty-four hours a day or just for certain times? The third item is to determine who will conduct the SD. Will you hire professionals or rely on family members and employees? The fourth issue is whether you will conduct the SD from your home, your workplace, or in between. Finally, you need to have a plan for how you will store and use the information you have collected. While you don't want to be in a panic, it is crucial that you are serious about this endeavor.

The most basic plan is focused on the needs of an individual, his family, or his workplace. His SD should accommodate specific needs of all those concerned with protection of the family and the home. By conducting SD from the house, it is possible to prevent attacks, a home invasion, or a kidnapping. The plan should focus on the transition times such as leaving for work and returning home at the end of the day. Before leaving the house, the individual can simply look out the window to see if anybody is watching. As he leaves, his wife can watch to see if anyone is following him. When he returns home, his wife can check to see if there is any suspicious activity taking place.

Development of an SD plan should involve identification of the best observation points in the house. Once you have identified the most likely routes for the arrival of trouble, you can match those locations to the windows used as observation points. As you are conducting this SD, it is important to be discreet so you do not scare off the surveillance. It is always to your advantage when the surveillant does not know of an SD. It is also important that your SD plan covers drop off and pickup times for any children in the family and creates contingencies for how to respond to an attack. This means that before anything happens, it is helpful to inform local law enforcement about this situation.

The next level of complexity for an SD plan is one designed for a small business. Like the individual SD, this one begins with a threat assessment. If the business has several buildings, each one will need to be studied as part of the threat assessment. The threats faced by a small business will vary according to the product being produced or marketed. If it is a controversial one, such as hunting supplies, you can predict which interests will threaten it. With modern technology, the small business can set up a monitoring device and designate certain individuals to be responsible for recording threat information. If the business is in a neighborhood plagued by crime, the threat may be a simple as shoplifting. The individual responsible for keeping track of the security system will probably be the owner, manager, or some other trusted individual. It is probable that a small business would not have the budget to bring in a specialist for this task. Like the individual who runs his SD out of his home, the small business can conduct his SD from his premises. Something as basic as security cameras can cover most of the needs of the establishment.

The small business SD is more complex than the individual's SD, but the corporate SD can draw from an abundance of resources. The assets of a corporation can easily run into many millions of dollars or even more. In addition to property, a corporation is likely to have a large work force that will also have interests that must be protected. Therefore, the corporation has a lot to protect and is obligated to institute an elaborate and detailed SD plan. The corporate plan will require the designation of specific individuals within the chain of command to be responsible for maintenance of SD procedures and will employ specialists with experience in this field. The specialists will take the lead in establishing comprehensive, complex arrangements to safeguard property and employees of the corporation. Because insurance for protection against terrorist acts is so costly if it's even available, it is better for the corporation to develop programs that offer a realistic prospect of protecting the corporation against such attacks.

CHAPTER 6

Hostile Target Surveillance

THERE ARE COUNTLESS TYPES OF ATTACKS based on targets, weapons used, and other considerations. There are, however, certain similarities regardless of the specific type of attack. The similarities are based on what is known as the Hostile Events Attack Cycle. Whether the attack is mounted by a sophisticated, organized group or by a lone individual, there are steps that will always be followed, and surveillance is their most common denominator.

As attackers plan, prepare, and execute their operation, the Hostile Events Attack Cycle demonstrates how they can succeed and often escape. While there are examples of incidents that were impulsive, most attacks require considerable planning that will often span weeks, months, or even years.

Apparently, Osama bin Laden began thinking about what became the September 11 plot in 1996. It was not until early in 1999 that bin Laden gave his approval for Khalid Sheikh Mohammed to begin formal planning. What followed throughout the year was a series of meetings about selecting individuals to manage the attack with special consideration given to the nationalities of attackers so they could more easily get visas to go to the United States. An al-Qaeda "military committee" was formed and led by Khalid Sheikh Mohammed that was responsible for the complicated operational support. This included selecting targets and arranging for the hijackers' travel to the United States. The list of potential targets was long and included nuclear facilities, but was finally cut down to four targets on the east coast.

In order to avoid having all of the participants located in the same place, the al-Qaeda leadership created the Hamburg cell in 1998. Mohamed Atta, who had gone to Hamburg to study urban planning in 1992, was in charge of the Hamburg cell. He was later joined by Ramzi bin al-Shibh who arrived in Hamburg in 1997. The two became roommates and both received training at al-Qaeda training camps in Afghanistan. Originally, bin al-Shibh was supposed to be a hijacker pilot but failed to get a visa to come to the United States. The selection of the hijackers was made by bin Laden and his lieutenant Mohammed Atef after a series of meetings with Mohamed Atta.

Before applying for US visas, the Hamburg-based terrorists were able to get new passports after claiming their old ones had been lost. Once group members were back in Germany, they followed guidance on proper demeanor by changing their appearance, as well as their behavior. They stopped attending radical mosques where they had met and no longer engaged in radical discussions. Their goal was to appear as moderate and as normal as possible in every way.

In order to support their visa applications, group members paid tuition for programs in schools such as the Florida Flight Training Center. Such applications were an essential part of the story they were telling consular officers at the US Embassy. Other necessary steps involved the creation of bank accounts and the transfer of money into those accounts so they could prove they had the resources to support themselves in the United States. The next step in this complicated effort was the hijackers' applications to attend airline training programs. Mohamed Atta was responsible for this aspect of their plan and sent over fifty emails to various flight training schools in the United States. With this success, in May 2000, Atta applied for and received his visa, as did the other members of the team.

There were considerable financial resources required for the US planning stage of the plot. According to the *9/11 Commission Report*, Atta distributed as much as $500,000 to the group members shortly before the day of the attack. The precise identities of those who had provided this support were never made public, but numerous official statements indicated a complicated network that led to Saudi Arabia. The *9/11 Commission Report* also paints a picture of failure by the FBI, the CIA, and other security services in the United States that failed in their investigatory and surveillance responsibilities.

It is easy to understand that a massive plot such as the 9/11 attacks would be complicated and expensive. However, the basic steps in a plot such as this and a more modest one involving a lone attacker are similar. In June of 2015, a white supremacist neo-Nazi named Dylann Roof entered a Bible study at Emanuel African Methodist Episcopal Church in South Carolina. In planning this attack, Roof followed the principles outlined in the Hostile Events Attack Cycle. He began his planning at least nine months before the actual attack. He visited the target on several occasions and even attended a church service there at one time. His preliminary observations were followed by a series of intense observations to determine attendance patterns. He rigorously practiced using his weapons, including rapid reloading procedures, and his escape plan. He even operated his own website in which he proclaimed his hatred of African Americans and other minorities. He outlined his motives, saying that he hoped to spark a race war in the United States and that he believed at the end of that war he would be pardoned by the victorious whites. He expressed his confidence that he would then be appointed as governor of South Carolina. Instead, he was convicted of murder and sentenced to death.

Whether the threat is complex, such as the September 11 attacks, or limited to one person, as in the South Carolina church attack in 2015, protection professionals require highly specialized training. An important step in this is known as Hostile Environment Surveillance Operations or HESO. To understand conditions affecting surveillance activities in a hostile environment, it is important to know what constitutes a hostile environment. Preliminary observations will enable a surveillance team to know if the region is under a threat of war, civil unrest, or insurrection. Less devastating but still critical to surveillance operations are high levels of crime or public disorder. The same impact on operations will be caused by natural events, such as earthquakes, floods, or hurricanes.

There is a clear distinction between hostile locations caused by environmental factors as opposed to man-made factors. This distinction does change the operational demands for surveillance teams. Activities such as covert filming pose special threats because there is often a popular, as well as an official, distrust of operatives doing this. The same risk exists for surveillants needing to observe terrorist incidents. For a surveillance team conducting a risk assessment during a reconnaissance mission, it will be important to report issues, hazards, and controls affecting future operations in that region.

The most important objective for Hostile Environment Surveillance Operations (HESO) is to gather information about any force that might constitute a threat and to use that information to prevent an attack. This is relevant for the protective efforts of a government or for an individual. In order to satisfy these objectives, HESO training is divided into four categories.

The first is protective surveillance (PST) which is an effort to create a protective ring around the potential target. Members of this team must identify any land or location that could be used as a perch for surveillance of the target. A team member must secure each such location whether it is a rooftop or a ditch. If any member identifies a hostile surveillance team, the close protection team must be called. Should their response not be quick enough, the protective surveillance personnel must act as a quick reaction force and protect the target or remove the target from the danger zone. For this reason, protective surveillance team members must be trained in tactics for close protection and kidnap prevention.

The second category is surveillance detection and is aimed at detecting hostile or suspicious activity before an attack can be mounted. The surveillance detection team (SDT) must identify any locations that could be used for observing the possible target they are protecting. By considering how they would attack their own facility and studying avenues of attack, they can develop a comprehensive detection plan. They will monitor the critical locations and watch for any traffic in those areas, including both foot and automobile traffic. The key is to determine which traffic consists of people or vehicles that seem out of place. The SDT is responsible for examining the travel routes taken by the target and when he would be most vulnerable to an attack. This work must be done discreetly so the hostile surveillance team does not realize the SDT is observing them. This situation will make it easier for the hostile team to become over-confident and make mistakes. If the surveillance detection team collects sufficient intelligence, it will be possible to call in counter surveillance personnel to organize an operation against the hostile surveillance team.

The third category in Hostile Environment Surveillance Operations is counter surveillance. CS includes actions taken to neutralize hostile surveillance operations. By observing the actions of a hostile surveillance team, CS personnel can develop deceptive efforts to mislead the hostile threat and exploit any mistakes they might make. A typical tactic employed in this context is to create alternative routes that the person you are protecting may take to travel from one place to another. The use of a multi-car convoy was used by the US-trained team protecting Augusto Pinochet when Chilean terrorists attempted to assassinate him. The chauffeurs employed defensive driving techniques and confused the attackers who guessed wrong about which car Pinochet was traveling in. Because civilian contractors are playing a bigger role in counter surveillance, these necessary skills are now on the market for would-be attackers to employ against protective services.

The collection of detailed intelligence that identifies the vulnerabilities of a potential target is the final category in Hostile Environment Surveillance Operations. This is referred to as Close Target Reconnaissance (CTR). If CTR personnel are protecting commercial buildings or residences, they will assemble intelligence that helps them identify possible entry points into the building and determine structural vulnerabilities of those facilities. This is work requiring a different skill set from those working in other aspects of HESO. Special emphasis is placed on a vast array of disciplines including environmental and terrain analysis, tactical tracking operations, covert patrol tactics, and facility vulnerability assessments. There are additional specializations that constitute the core curriculum of HESO training: advanced level training in covert physical surveillance, specialized training in night surveillance operations, target identification, and firearms training. Moreover, experiences in Afghanistan and Iraq have recognized the development of skills to detect suicide bombers and IEDs.

Among the most important facilities requiring protection are US embassies. The embassy is one of the most convenient targets for radical groups hoping to express their hatred of the United States. The history of embassy attacks can be traced back more than fifty years with 1968 marking the first chapter in this terrible history. At that time, an anti-Vietnam War demonstration in front of the US Embassy in London turned violent when several thousand demonstrators invaded the embassy grounds. Other attacks followed, most painfully, in 1979 when the US Embassy in Tehran was occupied by rioters who took the embassy personnel hostage and held them for 444 days. American embassies and consulates have become primary terrorist targets in recent years. In 1998, al-Qaeda gained infamy for attacking two US embassies in Africa on the same day.

Specific responsibility for the protection of embassies and embassy personnel rests with the Diplomatic Security Service. One recent innovation has been the creation of a safe room in most embassies located in what are regarded as hostile environments. CTR teams may be employed for special purposes such as management of convoys, their protection against terrorist attacks, and protection of employees in the safe rooms.

Diplomatic personnel are frequent targets, such as in 2012 with the attack on the US consulate in Benghazi, Libya and the deaths of the US ambassador and multiple security personnel. Over the last five years, dozens of American diplomats and intelligence officers have experienced a range of strange and often severe ailments. Most of these victims in these "health attacks" were serving in Cuba, Russia, or China although some were in the United States when struck by health problems, including traumatic brain injuries, part of something now known as "Havana Syndrome." There have been documented cases of at least twenty-one US victims who have suffered from long-term hearing loss and speech problems. In 2021, there was increased speculation in the media about this phenomenon because of a work injury lawsuit that was filed by a retired National Security Agency counterintelligence officer. The officer developed a rare form of Parkinson's disease that, according to his lawyer, was caused by some sort of high-tech weapon.

Apparently, this has been an issue within the US intelligence community for decades. "Havana Syndrome" victims experience nausea, ringing ears, vertigo, and traumatic brain injury. These reports prompted both the White House and the Senate Intelligence Committee to direct research on the possibility that high-tech weapons are being used to attack diplomatic personnel. Members of the Senate Intelligence Committee suggested that a high-powered microwave weapon may have been used in the attacks. This is a weapon that can weaken or even kill an enemy over a long period of time while leaving no evidence. In 2012, intelligence information indicated that this weapon was intended to flood the target's living quarters with microwaves. This would result in physical injuries such as a damaged nervous system. Working with bipartisan support, the Biden White House initiated a program to discover the cause of the "Havana Syndrome."

Russia is known to have experience with radio frequency research, and this has fueled speculation that the diplomats may have been attacked with a microwave weapon, as cited above. In the case of the injuries sustained at the US Embassy in Havana, there is suspicion that a botched surveillance job may have been responsible. The fact that attacks seem to have been limited to specific rooms or even certain parts of those rooms has added to official confusion about the attacks. Not surprisingly, this uncertainty makes the work of a target reconnaissance team more difficult. These problems seem to have been caused by "pulsed" and "directed frequencies" that are often associated with hostile surveillance efforts.

CHAPTER 7

Public Health Surveillance

A NOTHER ASPECT OF HOSTILE SURVEILLANCE THAT may lack the "romance" of surveillance against a terrorist team is the increasingly important area of public health surveillance. With the appearance of COVID-19 as a major factor in almost every aspect of national security there has been an understandable search for the data or intelligence about public health threats. This is the information that determines what requirements are imposed on businesses and individuals in resisting the threat of this new ailment. It also shapes public policy and the messaging that is used for encouraging compliance with those policies.

There are four basic aspects of surveillance associated with health problems. The categories are (1) infectious diseases such as COVID-19, (2) noninfectious diseases and health conditions, (3) risk factors, and (4) exposures. In recognition of these concerns, there has been an increase in surveillance-related grants and funding. The area for this surveillance effort is in clinical laboratories and the work is conducted by health scientists and medical doctors.

The starting point for this surveillance is notifiable diseases legislation. This legislation provides guidance about which diseases and conditions must be monitored. This surveillance constitutes the basis for reports that are issued and directed to agencies and entities that must respond. Clinical and laboratory-based surveillance is essential for determining the allocation of resources used to react to a health threat. It will also shape population-level prevention strategies that will be directed on a community basis. Because of the stigma attached to certain diseases, reporting on them may be done on an anonymous basis. Sexually transmitted diseases have long fallen into this category but, with the intense public fear of COVID-19, people who tested positive were ostracized by others. Public health laws identify the diseases that must be reported and designate the agencies with which their reports must be filed.

Surveillance strategies must take into account the privacy needs of individuals who could face discrimination if their health conditions became known. Those strategies must recognize their legal requirements regarding the privacy and confidentiality of medical data. Because of the frequent need for rapid response, clinical surveillance is needed to provide early warning about outbreaks of infectious diseases. The primary care provider who recognizes the signs and symptoms of a disease is the individual who will first recognize the appearance of a health threat.

Operating on the basis of the "notifiable diseases" list, the primary care provider will use whatever diagnostic aids are available to fulfill his surveillance requirements. This list is mandated by law, and it will change over time. The most likely diseases to be monitored are infectious diseases that can quickly spread, especially if that spread is facilitated by water, food, mosquitoes, or any other common vectors. Health-care workers are also concerned about any illness that can be prevented by vaccination and want to insure a rapid response to that need. They are also alert to identify sexually transmitted diseases, foodborne or waterborne illnesses, contagious diseases caused by airborne particles, and those like rabies or malaria that are transmitted by parasites. By 2015, there were requirements to report medical information in electronic claims data rather than in the previously less systematic fashion.

CHAPTER 8

Mobile Surveillance

UNLIKE PUBLIC HEALTH SURVEILLANCE, WHICH ATTRACTS attention during an era focused on COVID-19, the art of mobile surveillance is traditional and essential to the basic work of intelligence tradecraft or detective work. Mobile surveillance has complexities that are absent in the more sedentary static surveillance. Since we cannot count on a target remaining in one place, unless it is an embassy or some stationary structure, it is important to understand how to conduct surveillance on the move.

There are three categories of mobile surveillance. The first is surveillance on foot while the second involves use of one or multiple vehicles. The third is an amalgam of foot, vehicle, and public transport. If you think about them, you will recognize that they represent very different challenges and have unique strengths and weaknesses. However, what they have in common is that they are both based on an effort to remain invisible while being on a sidewalk or on a public road. Success in a surveillance operation is dependent on good planning. This means you need to be prepared to go from vehicle surveillance to foot surveillance without delay. If you realize you will be on foot, it is crucial to have clothing that is weather appropriate and will blend in with your surroundings. You need clothing that will enable you to change your appearance so your profile can be different from time to time.

An important step in planning is to conduct a walk-through of the locations that might be used in your operation. This includes identification of access points, side streets, and useful vantage points. The walk-through will be used to identify routes that might be taken by all those involved in the operation. It should cover actions to take if part of the plan does not work as expected.

Mobile surveillance can be conducted against an individual who may be hostile or it may be conducted where you are providing protection for a person. There will be tactics or requirements that are associated with each type of conduct. The overall general concern is that you do not want to be seen as correlating to the target by stopping whenever he stops and moving when he moves. This is true when the surveillance mission is to protect the subject or to observe the activities of a hostile surveillance team. The time of a target's transition from static to mobile is crucial. It is managed by calculating how long the surveillant can wait for his transition from static to mobile.

The first step in surveillance against an adversary is to study the location at which surveillance will begin. You will want to study the site and check for security cameras or any people who might notice what you are doing. It is important to know the local transportation system, identify the exits, and select a perch from which you can make observations. By surveying the entire area, you should consider how you will follow your subject should he decide to go mobile. Observe the flow of pedestrians and decide if there is enough traffic that you can closely follow your target when he moves. You want to be visible while in plain sight. If your actions fit your cover, there is a better chance of accomplishing this. When your target stops, you don't want to just stand on the sidewalk. If there is a convenient coffee stand, buying a cup will help you fit in. By looking ahead for plausible stop locations, it will be easier to blend into the area. In all of this, it is important that you do not make sudden or unnecessary moves that could attract the attention of a casual watcher. In a similar fashion, when your subject gets up to leave, wait a few seconds before you move to follow him. This helps prevent the appearance that you are following him.

When your target begins to move, don't follow him too closely, but do not allow him to get too far ahead of you. If he is too far ahead, you may lose him in the crowd. However, if you are following at a greater distance, it will be easier to adjust if the target happens to stop suddenly. If you are too close and he happens to turn around, he is more likely to notice you following his path. You can reduce this threat if you remain in his blind side behind other pedestrians. A significant danger point is if your target turns a corner and then stops to see if someone is following. Should you be too close to him, this move can ensure your discovery.

If your target moves into a subway, there is an increased risk that you can lose him. Because there are multiple destinations and stops, your target may get off very suddenly. This means you must be closer to him. You can also use windows or other reflective surfaces to observe your subject. This will help you avoid making eye contact with your subject who might then recognize that you are surveilling him. An alert subject might enter a phone booth, something that can still be found in train or bus stations with diminishing frequency, in an effort to pick out a surveillant. Since he might actually make a call, it could be useful to enter an adjacent booth. Even if you cannot hear what he says, you can try to see what page he might be on in a telephone book. With the prevalence of cell phones, this maneuver is less likely to be seen today. If your surveillance-wary target should stop and go back from where he came, do not immediately reverse. Continue as you were and try to reverse as soon as possible.

There will always be speculation about how many surveillants you should use. For maximum flexibility by members of the surveillance team, six is a workable number. If there are more team members, the target is more likely to spot the surveillance. There is a general assumption that three is the optimal number of members in a team. Three will provide reasonable flexibility and if your target changes directions, the lead operative can ignore that change and make no changes in his behavior. The second operative will then assume a position behind the target and continue the operation. This maneuver is referred to as leapfrogging. Utilization of a team also facilitates another maneuver known as the "floating box" in which team members surround the target on four sides. They will be in front as well as behind and also on parallel streets. As the target moves, the box moves. By shifting positions, there is less chance that the target will recognize surveillants.

One reason that three is the optimal number is that with three people you can place one member on the opposite side of the street where he is less likely to be seen but can still have a comprehensive view of the terrain. It will be easier for him to determine where the target is headed and the operative is less likely to be seen. One-man surveillance is not effective for mobile surveillance but it works very well for stationary observation of a target.

CHAPTER 9

Hostile Surveillance Detection

THE KEY CHARACTERISTIC OF HOSTILE SURVEILLANCE detection is that it is proactive. Its goal is to determine if someone has placed your client under surveillance in order to make an attack. While other techniques emphasize rapid responses to a threat, a hostile surveillance detection team will identify groups or individuals who represent a future threat.

Their first step is to determine if hostile surveillance is being conducted and to identify the responsible entity. This is accomplished by understanding what needs to be detected about other actors. How do you distinguish a random person on the street from the criminal or terrorist planning on kidnapping your client? Success in this effort will be served as the surveillance detection team members select the appropriate locations to observe their environment.

There is a common assumption that you can pick out the threatening individuals by noting suspicious behavior on people who did not fit in with the environment. While these can be valid indicators they are associated with non-professionals rather than serious teams that represent a real threat to your protected target. What matters most is correlation to the target. If your protectee moves and an individual in the area moves at the same time, this demonstrates his relationship to your protectee. It is for this reason that surveillance training stresses that you should delay, rather than immediately follow, your target. Any form of observation or communication in conjunction with the target will demonstrate correlation. Your goal is to blur correlation because it can never be completely eliminated unless you completely lose your target. At a minimum, merely observing a target creates correlation, albeit a rather small one. If you get up and follow the target, that would be a more obvious correlation.

There are additional aspects of correlation. One is referred to as correlation over time. If the target is staying at a certain hotel for several days and the operative makes it a point to watch in the lobby each day, that is correlation over time. He doesn't actually do anything; he is just making sure his target is still in a particular location. If his target is traveling to other cities and the operative goes to those cities to observe, that is correlation over time and distance.

A person conducting surveillance detection will need to see both the operative and the target at the same time to be certain if surveillance is taking place. A common vantage point for surveillance might be a bench in front of the building from which the target operates. Effective surveillance detection requires your presence not on that bench but at a vantage point from which the bench is observed. By observing an operative occupying the vantage point during times the target is active in or around the building, a surveillance detection team can make an accurate assessment of what is happening.

When dealing with professionals, surveillance indicators are usually very subtle. They may be actions which by themselves would have no significance. A member of a hostile surveillance team may use his cell phone to make a note, send a text, or even take a picture. Viewed in isolation, these are mundane acts. The surveillance detection operative can catalog these actions only if he is able to notice them and if he has the capability to note which individuals appear at the locations he is observing. This challenge can be met by, first, identifying the vantage points for each location and, second, by noting the individuals who occupy them. If any of those individuals appear more than once or twice, this will demonstrate that a correlation is taking place.

An individual who was assigned to the office of the defense attaché in the US Embassy in Prague during the Cold War spent his first free day in the city by taking a walking tour. During this time, he took many pictures. It was only later that he observed the same individual in three of those photographs. While his supervisor in the embassy warned him that he would be under surveillance, he gave little thought to this during his free time. In mentioning this later, he was informed that one appearance by the suspicious looking individual was chance, the second appearance was suspicious, and the third appearance constituted proof of surveillance by local security services. It is important to be aware of such activities but it is equally important not to imagine surveillance when none is present. The "rule of three" is generally accepted as a valid indication of the reality of surveillance.

As the hostile surveillance detection team makes these observations, they should ask what it is that the hostile team hopes to accomplish. An excellent examination of this issue was seen in 1978, when former Italian Prime Minister Aldo Moro was kidnapped. He was being driven to work in a two-car motorcade accompanied by five security guards. In planning for this attack, the terrorists conducted surveillance to identify the strengths and weakness of Moro's security. This enabled the terrorists to evaluate the opportunities for kidnapping the former prime minister as well as the threats they would face.

A brief summary of items that the terrorists studied prior to the attack demonstrates how easy it is to acquire simple information that ensures a successful attack. In very general terms they studied Moro's pattern of life. Because Moro's protection team used the same route almost every day, it was possible for the hostile surveillance team to study not only the route but also convoy management. What they saw was that the drivers in this two-vehicle convoy drove too close to each other and tended to tailgate other vehicles as they drove to his office. They studied the location selected for the attack and learned how to blend in. It was a neighborhood in which many airline personnel resided, so four of the terrorists wore Alitalia uniforms as they stood on the roadside apparently waiting for a bus. When terrorists noticed that a flower vendor parked his truck in what was to be the attack site, they slashed his tires that morning so he could not be there and cause complications for their plan. Telephone lines were damaged to create difficulties for anyone phoning for help. Finally, they had escape vehicles in place so they could escape after the attack in which eighty or ninety rounds were fired into Moro's convoy and all five of his security officers were killed. A total of eleven terrorists were involved in this short operation for which there had been months of planning by the hostile surveillance team. Their surveillance made it possible to spot the weaknesses in Moro's security and to take advantage of those weaknesses.

An intelligence collection plan must have specific objectives about the types of information needed. The Moro surveillance team wanted to know about the security used by the former prime minister and if they were well trained. Did they vary their protective operations or always use the same plan? Another important factor was to study the types of vehicles used by the Moro security detail. An analysis of convoy security was essential in the effort to identify vulnerabilities. Finally, they watched the Moro household staff to determine if any of them could be enlisted to help.

The performance of the hostile surveillance group will be a function of their level of training. Terrorist training did not become well organized until the 1960s when the Soviet Union began to support groups that could further Soviet worldwide objectives. The European Marxist terrorist groups that were trained by the KGB and East German Stasi were capable of very professional surveillance tradecraft.

The best-known location for this training was the Patrice Lumumba University in Moscow. It was founded in February 1960 as the People's Friendship University of Russia, and in February 1961 it was renamed in honor of Patrice Lumumba, an African leader recognized for his efforts to promote communism in Africa. Ostensibly organized to support international friendship, the university was known for recruiting and training terrorists throughout the Third World and training them in skills of espionage and terrorism. East Germany's Stasi and the KGB worked together to provide money, training, and opportunities for members of terrorist groups to develop their skills. There were major training facilities in Dresden and Karl Marx Stadt in East Germany. Soviet training was a global operation and was supplemented by the services of the Romanian Securitate and Cuba's General Intelligence Directorate.

The knowledge that group acquired was shared worldwide so by the time the Soviet Union collapsed, there were already training camps around the world. With the loss of Soviet support, such sophisticated surveillance became the exception, rather than the norm. As a result of this change, terrorist training was organized on the basis of regional, ethnic, and religious lines. Terrorist weapons became increasingly sophisticated and, as a result, the demand for extensive education became a factor in terrorist operations. Amateurs could not effectively use something like a mercury tilt fuse for a car bomb without specialized training.

Disparities in training and skill levels mean that many groups are not capable of conducting surveillance operations that will generate the needed intelligence. If a group is not assertive enough, they will fail to get important information. If it is too aggressive, it risks being detected. These limited factors can make it easier for a security team to identify hostile surveillance. However, the dedication of terrorist groups means that they can take all the time needed to slowly compile information about their target.

In the end, the skills and resources of the hostile surveillance group is balanced against the skills and resources of the anti-surveillance team. If there is a major disparity in their training, the best trained team is likely to win. The counter surveillance team must stress being covert so the hostile surveillance team does not realize it is present. If it has the skills to remain invisible, it will prevail.

Should the hostile surveillance team fail to observe the counter measures employed against it, they are more likely to make mistakes. On the other hand, if it is aware that it has been observed, its behavior may become erratic. In 2005, terrorists planning a raid on Nalchik in the Kabardino-Balkar Republic realized they had been compromised. Rather than back off, the group launched their attack prematurely and 142 people were killed. If Russian security forces had been sufficiently covert, they could have moved in such a way as to prevent such a devastating assault.

The extensive, well-organized training of security forces provides a degree of comfort. They, however, are held to a higher standard of training. A terrorist group may have two or three members that are skilled, but most of their members are not so well trained. Moreover, they can be selective about their targets and avoid those that are too well protected. The group that killed former Prime Minister Moro initially considered a different Italian politician, but backed off when their preliminary surveillance showed he was much better protected than Moro. Also, the less well-trained terrorist group has no concern about preventing collateral damage and often delights in greater carnage.

Each group will have a core of skilled people who take care of difficult tasks. The loss of these individuals can often destroy a group. This was the situation for the Provisional Irish Republican Army, which sent its less advanced members to do menial tasks such as placing a bomb at the target. This was a cause of problems when the bomb carriers accidently detonated the bombs prematurely. To alleviate this difficulty, the bomb makers began to install safety devices that had to be removed before the bomb would detonate. That, in turn, created another issue because nervous bomb carriers would forget to remove the safety device so the bomb would not detonate.

CHAPTER 10

Mobile Surveillance Detection

A S NOTED IN CHAPTER 6, MOBILE surveillance is more complex than sedentary static surveillance. That chapter noted the contrast between surveillance on foot and surveillance using a vehicle. This chapter is limited to the more basic distinction between surveillance of a static target and surveillance of a mobile target. Therefore, we are considering mobile targets and what can be done to determine if the mobile targets you are protecting are subject to mobile surveillance.

For purposes of this discussion, we are focused on the perspective of a surveillance detection operator who is responsible for his "mobile principal" or protected client. His job is to be aware of his "mobile principal" and other actors who have a correlation with him. The SD operative must watch for anyone correlating to his principal in terms of observation, movements, action, or simply presence by being near him.

The starting point is to think about your protected client and decide where you would position yourself if you were surveilling him. Those are the locations that will be central to your protective activities. In mobile surveillance detection, the most likely location for a hostile team is directly behind your principal and/or on the other side of the street. This will usually give him the best observation of the target.

Another pivotal point is any place where your principal changes the direction of his movement. The giveaway is when a nearby person makes the same change of direction. In order to expose the hostile surveillance operative, you can arrange to have your principal make a series of sudden moves in another direction.

It is important to make note of the distance between the surveillant and the target and also the speed with which the surveillant is moving. The distance and the speed is generally a function of (1) how fast or how slowly the principal is moving, (2) whether the environment is quiet and calm or not, and (3) whether the principal is moving among a static crowd or one that is moving around. What the SD operative should watch for is the presence of an individual who carefully maintains his distance from the principal. If the principal is going down a street that is not heavily trafficked, you may see a person about half a block behind him. What does that individual do if the sidewalk traffic is suddenly very crowded? If the individual moves to be closer to the principal, that is an indication of his correlation to the principal. If there is another speed and distance change as the principal nears a possible turning point, that is another key indication of a correlation between the surveillant and the principal. By this time, you can be sure that your client is being followed.

A person who is proceeding down a street will normally have to make certain short stops, such as waiting for a light to change. As this happens, the surveillant will usually be forced to stop, thus demonstrating his correlation to the principal. If there is a second surveillant just behind the first one, the correlation will be less obvious because the second person moves into the position previously occupied by the first surveillant. This is something that demonstrates the value of having surveillance conducted by a team rather than just one individual.

The advantages of a team are also apparent for the SD operatives. When the mobile principal stops, the operational area or "Red Zone" being occupied by the surveillance team can be expanded. As their operatives spread out, the operational area is larger. This gives the SD operatives a larger area to watch and increases the possibility they might miss a correlation.

As the mobile principal moves, there will be certain locations or events that are part of his movement plan. These could include attending a meeting or an event, having a meal, or returning home. Such events constitute a long stop in contrast to the unscheduled short stops for traffic requirements, which cannot be anticipated with any precision. When the SD team knows about this, it can position itself where it will be better concealed and essentially invisible to the hostile surveillance operatives. With this advantage, the SD operatives will know about traffic patterns or problems as well as key vantage points they should occupy.

If an unskilled surveillant stumbles into an environment that he realizes is covered, he may display his alarm, thus exposing himself. Of course, a skilled surveillant will know how to recover from this difficulty and continue to behave in a non-suspicious manner. If no person exhibits suspicious indications, the SD operatives must look carefully at everyone in the area. It is possible that a person that did not stand out at this point might be seen later at a different location strolling within sight of the "mobile principal." The SD team must occupy a location that enables them to watch the surveillant but makes it difficult for the surveillant to see them.

While there are similarities between surveillance detection and surveillance, there are important differences. The main difference is that for the SD team the surveillant is their target while the principal is the target for the surveillance team. Mobile SD operatives are responsible for monitoring the "Red Zone" as it moves or expands. Hopefully, the surveillants in the "Red Zone" will not even notice the SD team members. By watching both the principal and the "Red Zone" you will be able to identify the correlation.

This process is a complicated puzzle which is made more difficult because of the mobility of the process. Of course, the best technique for surveillance detection is by occupying a location and waiting for your target to show up. In order to occupy a static position, it is necessary to know what your principal is doing and where he will be when he does something. Remaining static is a possibility if the SD operative knows the locations at which the principal makes a change in his movements. The places at which this takes place become "choke points" at which the surveillant will have to be in closer pursuit and is likely to reveal his correlation to his target.

In order to make this possible, a Surveillance Detection Route (SDR), as noted above, is prepared by the SD operatives and given to the principal. If mobile surveillance is being used against the principal, the SDR will help expose it. The Surveillance Detection Route must include an assortment of short stops, long stops and other directional transitions. By observing these changes, the SD team will be able to spot the hostile surveillance operatives. It is important not to include too many transition points, because that could alert the surveillants to what is happening and cause them to break contact. When the SDR is over, the different SD team members will pool the observations they made from their respective vantage points in order to identify people whom they observed at different points. As they compile their observations they will create a more complete picture of potential correlations.

CHAPTER 11

Global Surveillance

THE TECHNIQUES EXAMINED SO FAR FIT into what should be regarded as tactical considerations. They relate to matters affecting individual, small companies, or corporations. What will be considered here are strategic issues that are associated with global surveillance. There are numerous companies with names like Global Surveillance Associates or Global Surveillance Systems. Such names are a tribute to a growing acceptance of the notion that being able to establish global surveillance is both necessary and inevitable. Thus, the term has become part of marketing strategies for enterprises devoted to facilitating universal surveillance of almost all human activity.

The origins of an international system for global surveillance can be traced back to the first years after World War II when the UK and the United States forged the UKUSA Agreement. Later, Canada, Australia, and New Zealand were invited to join them, and in 1971 they created a global surveillance network code-named ECHELON.

This became the focus of international attention in 2013 with the revelations of an NSA contractor, Edward Snowden. Snowden released an enormous cache of documents that he collected while working at the agency. Many of these documents were court orders, memos, and policy documents regarding surveillance activities. It was the Snowden disclosures that presented a picture of the larger network of global surveillance that included the communications of prominent leaders of the anti-Vietnam war movement. There were also stories about spying not only in the United States but also overseas. The names of secret surveillance systems such as PRISM and Tempora became household words. Coupled with controversies associated with the 1972 Watergate break-in, there was a storm of criticism focused on all intelligence organizations and activities.

The five members of the UKUSA Agreement together became known as the "Five Eyes." Their overall objective was to achieve total information awareness, something that could now be accomplished by use of the internet and a variety of analytical tools. The NSA's director, Keith B. Alexander, confirmed that the agency was collecting and storing all telephone records of every American citizen. The Utah Data Center (UDC) was created to help store the enormous quantity of data drawn from emails, phone calls, text messages, and cellphone location data. This included conversations from over a billion users from all around the world. Its storage capacity is counted in exabytes, a previously unknown quantity. The UDC is also known as the Intelligence Community Comprehensive National Cybersecurity Initiative Data Center.

While its precise mission is so secretive that nothing can be told about it, it does support the Comprehensive National Cybersecurity Initiative (CNCI). The USC is in Camp Williams near Bluffdale, Utah and its work is supervised by the NSA. Many of the programs used by UDC and its collaborative entities undertake data mining from central servers and internet backbones. There is no limit to what can be picked up and it is impossible to predict what agencies or nations will be encompassed by its global reach.

While UDC is part of the NSA network, other agencies such as the United States Department of Justice and the FBI are closely involved with its operations and can call upon its services as they pursue a variety of surveillance objectives. Information gathered through these programs is passed to law enforcement authorities in order to initiate criminal investigations of US citizens. By manipulating the sources employed, it is possible to hide the provenance of this information. In the past, this could not have been done. Through the top secret Venona intercepts program of the 1940s, information was gathered about the espionage work of Alger Hiss, Ted Hall, and others, but that information could not be used in prosecutions. Another unfortunate side effect of these innovations is that the so-called red line between domestic and foreign intelligence operations no longer has the same impact.

All necessary orders to this work are signed by the Foreign Intelligence Surveillance Act Court (FISA). Numerous foreign intelligence agencies, such as Britain's GCHQ, are part of these operations. Several telecom providers, including AT&T, Verizon, and BellSouth, routinely share information about their customers. In 2008, there were reports about the existence of a program known as the "Quantico circuit" through which the federal government was given access to a "backdoor" into the Verizon network.

Among the many documents released by Snowden, there were indications that these programs were less about deterring terrorist attacks than about acquiring data about business, economics, and stability of nations that might be rivals of the United States. Many critics complain that the NSA never prevented any terrorist attacks while NSA Director Michael Hayden insisted their surveillance programs had disrupted fifty-four terrorist attacks. Hayden also admitted that the NSA steals secrets not to make anyone rich, but to make everyone safe. These allegations and boasts lack the specificity that would provide a comprehensive answer to these questions but it is obvious that many of the efforts advanced economic espionage, diplomatic initiatives, and social control. During questioning in US Senate hearings, NSA officials have argued that it is crucial that all conversations by American citizens should be stored and be available at any time that information is needed for a federal investigation.

XKeyscore is an international surveillance tool used by intelligence analysts to go through massive databases in order to locate specific emails, online chats, or browsing history of almost anybody who is suspected of being a national security threat. Tempora, the UK's global surveillance program actually intercepts fiber-optic cables used by the internet. The NSA uses a surveillance program known as PRISM that can reach into the servers of Microsoft, Google, AOL, YouTube and others to examine messages that have already reached the party for which they were intended.

An objective of this massive high-tech operation is to ensure that the state can know the most important information about each person. This means being able to display that person's social connections, people with whom they may travel, the destinations to which they travel, their hobbies, or their reading preferences. This can all be accomplished by studying their phone calls, emails, or text messages. During the Cold War, the KGB, the Romanian Securitate, and other intelligence organizations would often require suspects to sit in a room for several days and write a complete autobiography charting their development from childhood on up to that very day. Often they would be required to do the autobiography a second or even a third time. The result would generally be a document at least three hundred pages long. With all of the data routinely gathered by the NSA, the computers can now provide a complete picture of our personal, professional, religious, and political affiliations without any active contribution from the subject.

There is an aspect of most investigations that is focused on financial matters. Referred to by the catchphrase "follow the money," this is based on an assumption that simply by determining who sends money to whom, you can understand what is happening. In recognition of this, the NSA created a branch known as "Follow the Money" (FTM). This branch has the capability to monitor all banking transactions to include international payment and credit card transactions. The NSA uses its financial data bank to store this data so it will be available for analysts.

Through FTM, the NSA can monitor the global flow of money and systematically gather important financial intelligence. With the collapse of the Soviet Union in 1991, illegal international arms traffic, already a major trade, expanded to include almost any weapons in the massive Soviet arms storehouse. A major hub in this burgeoning trade was the Kolbasna arms depot in the Dniester Moldovan Republic, the USSR's largest armament storehouse. Guarded at the front, the back gates were almost always open for shipments to global destinations. The only way to understand this weapons trade was by going through the financial transfers of the buyers and sellers.

Because of the social distancing associated with COVID restrictions, more companies are forced to do business over a digital network rather than in face-to-face meetings. Even traditional brick and mortar businesses have been forced to increase digital transactions as customers pay online and have the purchases delivered. Online purchases through Amazon and eBay are now the modern norm. Not surprisingly, digital skimming has become a major threat to all business establishments. The FTM system has also facilitated surveillance of cybercrime.

For every online transaction, the company has to secure and maintain personal data about the customer. This data has enormous value and is often marketed online as a commodity in its own right. Every person who has an account with Amazon has surrendered their bank card information and verification codes along with basic information such as names, physical addresses, email accounts, and phone numbers.

The actual theft takes place at the user's computer and it is possible that he will not realize what has happened for months. During this period, hackers have free use of credit and debit cards. This information can easily be sold on the dark web. Those who get this data are able to steal millions of dollars from users around the world.

Even a novice can use this technique in order to steal from unwary card holders. By going to the dark web, you can purchase toolkits that provide professional level training for digital skimming. A transaction begins when a user downloads a website for an e-commerce company as he prepares to make a purchase. If a hacker has penetrated the e-commerce site and entered a skimmer code, information entered by the customer automatically goes to the hacker, who can then sell that information. Hackers can create lookalike domain names to convince users they really are communicating with Norton Anti-Virus or even their own bank. Fake checkout pages can be created at the very end of the buying process in order to rob clients who believe they have completed their purchase. From the first key strokes, the hackers have stolen valuable information that will be used to rob naïve users. Sophisticated hackers often employ "automated bots" that can run thousands of transactions each second in an effort to find valid username/password combinations.

These actions often lead to identity theft. If you are a victim of identity theft someone else will use your personal identifiable information to gain financial advantage. This has involved the purchases of automobiles, personal luxury items, or even homes. Personal identifiable information can include your name, your date of birth, your social security number, and your driver's license number. If the hacker manages to get the user's electronic signatures, fingerprints, or passwords there is no limit to the amount of damage that a victim can incur. The individual victim does not always know that he has been targeted for identity theft. Many data breaches are massive, such as attacks on Facebook, but not all of the stolen data is used. The better your credit rating, the more likely it is that your information will be used.

When people steal identities in order to do things that are illegal, this is referred to as criminal identity theft. There have been numerous cases of people who manage to get copies of professional certification documents of doctors, lawyers, or other professionals. These are people who might have decided to skip medical school but practice medicine anyway. With the assortment of WebMd and other sites offering medical advice, it is not impossible to fake being a doctor. As long as you limited yourself to relatively simple cases and refer seriously ill patients to a specialist, your fraud might not be detected.

While professional hackers are most often associated with identity theft and are constantly involved in this activity, they are not the only ones to steal an identity. Some of the others are undocumented immigrants seeking to hide their status. There are also people who have massive debts and are hiding from collection agents, as well as people who just want to become anonymous or start a new life. Some of the people who do this are not motivated by avarice or any desire to steal money. Quite often they simply want to live another person's life. Known as posers, they will lurk in social networks, using the photograph a person they wish they looked like and creating stories that will help them gain acceptance by friends of the person they are impersonating.

CHAPTER 12

Technological Innovations

S PECULATION ABOUT THE RISE OF THE surveillance state tends to focus on modern technology and assumes this development took place in the United States. Without a doubt, creation of the surveillance state has been driven by technology, but it may have taken place earlier than most observers think and in a far more remote region than assumed. Some would suggest that surveillance began in the Philippines in 1898.

The United States experienced a technological renaissance in the 1870s when Thomas Edison introduced the quadruplex telegraph and Philo Remington began marketing the typewriter. The invention of the electrical tabulating machine and the Dewey Decimal System gave us the ability to catalog and retrieve data efficiently. Add to this the growing availability of modern photography, and it is possible to see how we could not manage and transmit information in a systematic manner.

By themselves, these inventions did not immediately translate into essential components of a governmental apparatus. However, when the United States occupied the Philippines in 1898, the stage was set for creation of a system that could be used to accomplish important political goals. The goal in the Philippines was the suppression of a Filipino resistance movement that had become a thorn in the side of US authorities. In order to advance this objective, under the direction of Captain Ralph Van Deman, the United States created an amazing apparatus that could provide detailed information on Filipino leaders suspected of working with the resistance. With modern photography, the physical appearance of each person was recorded. Something as basic as the typewriter meant recorded information could be easily read. Other devices made it possible to have a comprehensive record of personal finances, family connections, and political associations. Eventually, this system had records on about 70 percent of the population of Manila.

The legal framework supporting the use of this system was created by the Sedition Act, which spelled out severe penalties for anyone engaging in subversive activities. The Philippine system was so effective that it was eventually imposed on the United States when the Wilson Administration faced domestic opposition to involvement in World War I. The necessary legislation in the United States was provided by the 1917 Espionage Act and the 1918 Sedition Act.

The military intelligence division created by Captain Van Deman cooperated with the American Protective League. Together, they compiled a million-page catalogue of reports on German Americans as well as other people who opposed Wilson's policies. Over ten thousand people were arrested in nationwide raids led by Attorney General A. Mitchell Palmer and J. Edgar Hoover. In 1919 Herbert O. Yardley founded the Cipher Bureau, better known as the Black Chamber. Its primary mission was to spy on foreign communications, including those of American allies. When the American public learned that most of the targets of this system were US citizens, there were demands to limit domestic surveillance, the State Department's Cipher Bureau was abolished, and use of the Espionage and Sedition Acts was curtailed.

Today as in the past, new technologies are dramatically altering prospects for the ability of government or private sector actors to monitor activities of citizens and other governments. Generally, surveillance was conducted by an individual or a small group of operatives. With the passage of time, spyglasses were replaced by telescopes and eventually radio technology augmented the work of surveillance teams. More recently, CCTV, RFID, and GPS technologies have revolutionized surveillance.

Surveillance technologies make it possible for a variety of actors—both governmental and private—to violate not only the law but also social norms about the expectation of privacy. Employees of various security services have used their access to sophisticated devices to spy on wives or girlfriends. Several years ago, council employees in Liverpool, England used CCTV cameras to spy on a woman in her apartment. In Pennsylvania, parents of students took legal action against high school officials who accessed the webcams on students' laptops to observe their at-home behavior.

So many commonplace items now have cameras that it is difficult to determine who is being watched by whom and for what purposes. As civil libertarians have grown concerned about the loss of privacy, there has been a debate about whether the use of surveillance devices enhances public safety or creates an Orwellian society in which "big brother" is always watching. With the rise of social media and the apparent willingness of the tech giants to share information with governmental agencies, the expectation of privacy that existed in the past is increasingly diminished. When Facebook and Apple agreed to give government officials access to the personal information of their users, it became obvious that the division between governmental and private authority had been eroded.

Cell phones have become major targets of technological innovations that provide access to their data. Normally, law enforcement agencies need to secure a search warrant to gather such information. However, in an increasing number of cases, those agencies have found a way to avoid the search warrant requirement. They do this through their use of a simulated cell phone tower—known as a Stingray—that secretly gathers information from all the cell phones in its area of operation.

As we live in an era of technological innovations, it is possible for the range of surveillance to reach far beyond what was possible at the time of World War II. Studies of the accomplishments of the Central Intelligence Agency demonstrate the agency's embrace of technology. Allen Dulles and Richard Helms, both of whom headed the CIA, were responsible for an effort to use modern technology to advance intelligence work. This led to the creation of the Technical Services Staff in 1951. TSS was greatly expanded and renamed the Technical Services

Division in 1960. In many cases the resulting innovations supported general intelligence work, including new weapons. In other cases, the innovations made a remarkable difference in what could be accomplished through surveillance.

Not knowing what your enemy may be planning is what we think of as an intelligence failure. For years, memories of the Japanese attack on Pearl Harbor drove intelligence organizations to seek better ways to observe the enemy. Shortly after the end of World War II the sensitivity of American intelligence services was heightened by their failure to detect Soviet intentions to impose a blockade on all land routes to Berlin. Later the outbreak of the Korean War was another reminder of the inadequacies of Western intelligence organizations to anticipate the moves of their adversaries.

The CIA was smarting from its obvious failures and the British security services were feeling the embarrassment of the defections to the USSR of senior British MI-6 officials Burgess and Maclean when they informed the CIA of a major technological breakthrough in 1953. At this time, the CIA leadership believed that covert operations would be the key to success in efforts to manage effective surveillance of the Soviet Union. The British technological breakthrough raised the possibility of a massive wiretapping operation that would undermine Soviet operations. This technology would open a wide range of Soviet communications without endangering Western operatives.

William King Harvey, who was chief of the Berlin Operations Base, became the American leader in organizing what became known as the Berlin tunnel operation. Although he had reservations about working with the British, the project became a British-American collaborative effort. To their credit, the British had managed to dig a tunnel in Vienna that enabled them to tap into the phone lines that ran to the Soviet headquarters located in the Imperial Hotel. For several years they acquired sensitive information about Soviet operations. The CIA later did the same thing in Vienna and concluded that if they could tap into Soviet communications in Vienna, they could do the same in Berlin which was even more important as an intelligence target.

The formal starting date for the Berlin tunnel operation was December 1953. It became one of the most spectacular signals intelligence operations of the Cold War. The closest the Soviet communication lines came to the west was in Rudow, a remote suburb in the American sector of Berlin. The tunnel in Vienna was seventy feet long while the one in Berlin ran for 1,476 feet. The tunnel began in a warehouse built to look like a typical US logistical facility. Displaced dirt was stored in the warehouse. All of the work was conducted within sight of guard towers along the sector boundary. The tunnel went through sandy ground and a cemetery. Flooding as well as a nearby sewer line made construction of the tunnel more difficult. The digging was completed in February 1955 and it took another month to complete the tap chamber.

As soon as the cables were connected, collection began immediately. A small group of linguists worked at the site. Voice recordings were sent to London for translation and all telegraphic material went to Washington. The system worked for over eleven months until East German soldiers began digging at the site of the Soviet cables and an alarm sounded in the tunnel. Harvey realized what was happening and removed the most sensitive electronic equipment the day before the tunnel was penetrated on the Soviet side. The tunnel operation, which has always been regarded as a great success, was exposed by George Blake, a senior British intelligence official who was a Soviet spy.

Throughout much of the Cold War, the Soviets were well ahead of the United States in their surveillance techniques, so the Berlin tunnel operation came at a time when the West needed an intelligence victory. The Soviets had expected their exposure of the tunnel to be a major embarrassment to the United States but global opinion indicated that most people were impressed by the skill and audacity of the US-UK team.

The case of Oleg Penkovsky, an officer in Soviet military intelligence (GRU) who was spying for the West, represented another Western intelligence success. When the KGB became suspicious of Penkovsky, they mounted a very effective surveillance operation.

As Penkovsky communicated with his handlers, the KGB had three observation points from which they could monitor the spy. Audio surveillance was managed from the apartment directly above his while visual surveillance was conducted from the balcony outside his window. Finally, their third observation point was set up in an apartment in a building across the river from Penkovsky's residence. From these three locations, the KGB was able to compile a complete record of Penkovsky's activities.

Before any operation could take place, the CIA needed to detect and counter the KGB's surveillance of CIA personnel and assets. The KGB's advantage was based, in part, on their assumption that any Westerner in the USSR was a likely spy. For each individual, the KGB prepared expected activity charts and responded immediately to any deviation from that pattern. Any attempt to evade KGB surveillance would bring retaliation that was often brutal. Their surveillance extended to the US Embassy and in 1963 embassy security personnel discovered wires running through the mortar outside the building and going into the basement. Because of the proficiency of KGB surveillance operations, the CIA stopped most of its agent operation in the USSR.

Technical innovations gave the CIA the ability to conduct overhead reconnaissance through development of the Corona satellite in 1959. Eventually, recognizing that the satellite was not inhibited by the KGB's superior ground surveillance capabilities, there was a decision to rely more on technological collection of information. Imagery analysis gave the CIA near real-time images of not only Soviet territory but also the territory of almost any country in the world. Intelligence agencies appreciated that this could be accomplished without creating risks for agents like Penkovsky who was executed by his bosses within the KGB and the GRU. Moreover, "big technology" such as the Corona satellite operated in a predictable, relatively clean environment. This was an advantage not enjoyed by the small devices that would be used in an unpredictable environment affected by dampness, dust, and rain. Electronic gear suffered when used in normal outdoor conditions.

While technology helped the United States overcome deficiencies in its reconnaissance capabilities, it did not determine the intentions of the Soviet leadership. Looking at the position of troops on the ground does not reveal the actual intentions of your adversary. In the end, a reconnaissance satellite was no substitute for effective surveillance. Therefore, the CIA leadership was committed to the selective application of technology and the Technical Services Division was responsible for making this work. CIA recruitment then had to shift from prestigious liberal arts universities that had long supplied the ranks of America's espionage services.

As the TSD developed modern tools that could be used in surveillance operations, there was a problem because such devices would stand out in the Soviet Union where citizens did not have access to modern technological innovations. An instrument that could be used in the USSR would have to be disguised to look like a more primitive gadget to which Soviet citizens might have access. Even a battery powered transistor radio was beyond the reach of a Soviet consumer. So the TSD set out to create special cameras, communications equipment, and counter surveillance devices. Without these, it would not be possible for CIA assets to operate in the Soviet Union. The CIA technicians also needed to develop sophisticated concealments to hide the tools to be used in operations.

One of the most prominent architects of the CIA's technical services was Seymour Russell. Most people recognized that stealing information was a difficult challenge. However, many agents like Oleg Penkovsky were exposed when they tried to communicate that information to their handlers. Russell worked to develop covert communications because it was required for operations every aspect of intelligence activity in the Soviet Union.

Concealment was the first step in this process. In order to disguise film, documents, or anything of intelligence value, agents needed concealment devices. Those items might be picked up in the basement of the Kremlin, but unless they can be transported to a location controlled by the agency directing these efforts, they were of no intelligence value. An agent that is making that journey is a great risk from his enemy's surveillance teams.

There are numerous devices created for the transport of items stolen by spies. The Soviet spy known as Rudolf Abel used a hollow nickel to hid microfilm. It worked well until he used that nickel to pay his paper boy. Fake batteries, hollow hair brushes, and cans of shaving cream have been shaped to hold small items of intelligence value. In the popular spy comedy show *Get Smart*, a shoe was used to hide a secret telephone in the days long before cell phones. In fact, transmitters were often hidden inside what were otherwise functional shoes. The German Steinbeck watch supported surveillance teams by doubling as a camera while still working to tell time. A thermos was also modified to hold a camera for use by surveillance personnel. When the micro-chip replaced the vacuum tube, it was possible to dramatically reduce the size of the instruments used for surveillance.

Miniaturization became the watchword for creation of technology used by surveillance personnel. The most important objective was the development of a short-range agent communications device (SRAC). In 1975, the only existing SRAC had a limited number of characters. If an agent had several pages of data to transmit, this would require the sort of risky moves that invited challenges by KGB surveillance operatives. For the transmission of a document of several pages, the CIA technicians needed to develop a "bubble memory" system. This was a system that could store and retrieve large amounts of digital information in a tiny sphere.

Success in this technology required another innovation: a "charge-coupled device" or CCD. It started as a memory storage device that relied on a chip consisting of light sensitive capacitors. The detail in each transmitted image was a function of the number of capacitors or pixels. The CCD eventually became a "filmless camera" or what is now known as a digital camera. Using this innovation made it possible for relatively high-resolution images to be immediately transmitted. Had he enjoyed the benefit of the digital camera, Oleg Penkovsky could have eluded KGB surveillance that had targeted him. By reduc-ing the size of gear used by surveillance teams, it is easier to hide it from observations of surveillants. This reduced the power required for operation of the device and made it easier to carry.

As long as Soviet surveillance capabilities exceeded the counter surveillance skills of CIA operatives, it was virtually impossible for American operatives to function inside the USSR. If a Soviet official was to be approached, the approach would have to happen overseas. So the CIA's targets were Soviet personnel stationed outside the Soviet bloc. One of the CIA's most valuable assets, GRU General Dmitri Polyakov, was recruited in 1961 while serving in the United States. Even working in a neutral or a Western nation, there were severe limits. KGB surveillance teams monitored the activities of Soviet citizens when they were overseas. Living in Soviet compounds, they had to deal with the informer networks that existed in Soviet institutions no matter where they were located. The ambition of CIA operatives was to meet and recruit Soviet officials who were stationed overseas but would eventually be transferred home to Moscow. Hopefully, their new postings would be in agencies conducting work of interest to the West.

The best-known case of such a recruitment this took place in Columbia and involved Aleksander Ogorodnik, an economist working in the Soviet Embassy. His recruitment and his training placed stringent requirements on CIA personnel responsible for surveillance regarding Ogorodnik. Unlike Dmitri Polyakov and others who spied for the United States, Ogorodnik had no training at all in intelligence work. The OTS sent qualified personnel to Columbia in order to train him during the few times he was available. Those sessions required Ogorodnik to learn surveillance and counter surveillance skills. He had to master use of an ultra-miniature camera known as a T-100 and later, to use an easier version of the camera known as a T-50. Both were designed to be hidden in what looked like a luxury pen. Communications were a challenge but as the OTS was utilizing new technologies, face to face meetings could be avoided. There were times when Ogorodnik would appear to be talking to a tree because CIA microphones were hidden in its branches.

After four years of successful operations in Moscow by Ogorodnik, his case officer found him to be non-responsive to requests for a meeting. After a four-hour surveillance detection run, his case officer, who was using an OTS-developed frequency scanner to detect Soviet surveillance, was confronted by a KGB team. Within a few hours, the case officer was expelled from the Soviet Union. The previous month, Ogorodnik had been confronted by KGB officers in his home and used his suicide pill to avoid interrogation. From the KGB perspective, this constituted a defeat while the CIA regarded the four-year operation as a success.

Nonetheless, there was an eventual increase in clandestine operations by the CIA in the Soviet Union. The key to this increased pace was the deployment of new "spy devices." Agent communications and surveillance counter measures were two of the most significant technological improvements. Most spies had been tripped up when attempting to communicate the information they had stolen to their handlers, who could transfer the data back to the headquarters. A variety of short-range agent communications instruments helped reduce this danger. As noted elsewhere, surveillance counter measures are those actions to identify the presence of surveillance and to evade individuals involved in that effort. The key in this undertaking is to determine the objectives of the surveillants.

As impressed as we are with the emergence of new technological innovations, it is important to recognize that there are limits to the gift of technology. In the 1960s, the CIA ran a program that involved experiments on animals such as rats and ravens. The CIA's engineers hoped to use a tactic called passive concealment. Rats and ravens were noted for their abilities to get into places where they were not wanted, but cats were identified as being more appropriate for this operation. A power pack was placed in the abdomen of the selected cat and a cord was fitted next to his spine and connected to a recording device hidden in the cochlea of the cat's ear. The cat's tail became an antenna. These measures were intended to transform the cat into an instrument that could casually position itself near a target and record the target's conversation.

If successful, the project would have created an actual cyborg kitty. There were, however, some obvious design flaws. One of the first was the difficulty of ensuring the cat would follow directions. Cats, of course, are not known for their obedience. One of the targets of the acoustic cat was rumored to be an Indonesian politician who often met on the verandah of his residence located in a rural area. The area had a large population of stray cats so the official assumption was that another cat would not be noticed. He might not have been noticed by the people on the verandah, but other cats would have noticed this intruder. The ensuing confrontations would likely prevent the cat from assuming a position under the chair of the target. He probably would have been more attracted to possible sources of food in the area.

A second difficulty was the danger that the body of the cat could not indefinitely support the devices installed in his body. There were doubts that an altered cat could survive without constant attention from a veterinarian. Infections would always constitute a threat.

A final difficulty was the placement of a device to monitor conversations picked up by the cat. If used in a rural area near the home of a prominent politician, a van used for listening would have been observed and ordered to leave the area. The details of this failed project were finally released in 2001, when former executive assistant to the CIA Director, Jeffrey Richelson, published his book *The Wizards of Langley*. Even though the acoustic cat operation was an expensive failure, there have been similar efforts to transform insects into surveillance devices that could fly over the barriers protecting intelligence targets. Visitors to the CIA museum can see some of these innovations, such as a fish with a periscope, on display as tributes to the agency's technological skills.

CHAPTER 13

Challenges and Blessings of Technology

THOUGH THE IMPROVEMENTS WERE CONSIDERABLE, THERE were some problems associated with the new technologies. They were often a mixed blessing. The gadgets and the computers collect information. The Berlin Tunnel operation produced more data than could be processed by staff support elements. The CIA official responsible for the project joked about placing limits on how much information the Tunnel was allowed to send. When the eleven-month operation ended, the Tunnel data continued to be the subject of analysis for another two years. In short, the new technologies created a deluge of data meaning that analysts were tormented by having to determine which information was relevant and which was insignificant.

A basic objective of new indirect methods of communications often ran counter to the general impulses of agents who were risking their lives and felt that they deserved personal handling. Thus, their handlers needed to convince them that personal meetings might expose them to KGB surveillance. This was not always an easy task. With Adolf Tolkachev, a Soviet engineer who provided valuable intelligence about Soviet aircraft capabilities, the CIA made an exception. Tolkachev knew nothing of spy tradecraft and insisted on personal contact. This exception worked because, first, he was assigned a case officer who was a native Russian speaker and a master of disguise. He could look, smell, and act like a typical Russian. A second factor was the development of new technology that made it easier to detect Soviet surveillance. When Tolkachev was exposed in 1985, after almost six years of supplying sensitive information to the CIA, it was not the result of failed tradecraft but because the spy had been exposed by the CIA traitor Aldrich Ames.

The delivery of a sophisticated gadget and training the agent in its use emerged as another difficulty. Most Soviet citizens were products of an environment in which complicated devices were rare. Faced with such a novel instrument, they might become confused. Being unfamiliar with modern technology, it was likely that they might expose it to severe conditions or drop it. If that happened and the gadget stopped working, there was no convenient repair facility. Also, something that performed well in a sterile laboratory environment might not work well when exposed to moisture, dust, or mud. Even a tech savvy Westerner would have difficulty with these technologies. Finally, even though these new tools were fairly small, the agents would face challenges in finding a good hiding place. All of these creations were a sharp contrast with anything a Soviet citizen might possess.

The electronic short-range agent communication system (SRAC) was coming into its own during these years and was helping agents to communicate with each other. It measured three by six inches and worked as a "burst transmitter." The receiver device could be placed in a window and when the agent came by, perhaps in a bus, he could depress a button that sent a burst of information to the receiver. This was somewhat like sending a text message with a modern cell phone and made it possible for the agent to appear to be doing nothing unusual. It was easier than having to service a dead drop while worried if a surveillance team was watching. CIA technical operations officers used covert signal path surveys to identify the most favorable transmission points. Keeping the device charged was a challenge so the CIA engineers decided to use rechargeable batteries. This innovation led to another problem: finding a hiding place for the recharger unit.

In 1981, after five years of operational planning, the CIA launched one of its most elaborate surveillance operations in the agency's history. This operation linked ultra-sophisticated space age technology with traditional surveillance tradecraft. It demanded cooperation between a number of agencies including the CIA's Directorates of Operations and Science and Technology, as well as the NSA. Finally, the National Photographic Interpretation Center provided valuable support in planning for the operation known as CKTAW. The objective of the operation was to wiretap underground communications links between the USSR Ministry of Defense and a research institute in Troitsk, a closed city located twenty-three miles from the center of Moscow.

The first step in this operation was made possible by a new type of reconnaissance satellite, the KH-11 Kennan, launched by the National Reconnaissance Office in 1976. The KH-11 Kennan is a large vehicle that is sixty-five feet long and ten feet wide. Its most important component is a telescope that has a 2.4-meter-diameter mirror with sensors that enable detailed observation of almost anything on Earth. It can identify objects as small as ten centimeters wide. This satellite observed the Soviet military preparing a trench for communications cables between Moscow and Troitsk. Accomplishing what no on-the-ground surveillance team could have, the KH-11 noted that there were manholes along the length of the trench. The CIA aim was to gain access to the cable by using one of the manholes to enter the trench and attach a monitoring collar on the cable.

This act required a careful study of how the manholes were constructed and what was necessary to enter one without being exposed to KGB surveillance. Several surveillance trips were made to the area. Since visiting the location on a regular basis was dangerous the local CIA team was careful and took two years to complete their surveillance of the site.

This part of the CKTAW operation called for traditional surveillance tradecraft. The very small Tessina camera, which could be strapped to the surveillant's wrist, was used for casing photography. A specific manhole was chosen because there was a tree line that would provide cover for an agent approaching the manhole cover. Before the approach, agents had to determine how difficult it was to remove the cover. Next, they needed to know several things: (1) the precise measurements of the underground chamber; (2) the amount of ground water in the chamber; and (3) the accessibility of the cables. Not surprisingly, the cables were shielded in lead and protected by sensors and alarms. Once agents approached the cables, they needed to have the skills required for tapping into them. That task called for an expensive program to give agents those skills and required instruments. The most important of these was a collar that would surround the cable and record the signals.

And, of course, a preliminary examination of the cables had to be conducted in order to determine which cable carried the sort of information that would justify this expensive operation. This examination had to be conducted by a CIA officer disguised to look like an ordinary Russian out for a day in the sun. The OTS officer responsible for this aspect of the operation had been in Moscow long enough to establish a pattern of behavior that had been observed by KGB surveillance. It involved lots of family trips to parks and other nature settings. His behavior was not exceptional but rather boring. He simply did not look interesting.

The cover for his entry into the manhole was a typical family picnic. He and his wife wore radio monitors set to pick up possible KGB surveillance activity in their area. Once certain that there was no surveillance, he pulled his car into the spot they had selected. He avoided any locations typically used by Americans because he did not want to stumble into someone else's surveillance. He left the family picnic wearing a back pack. At each stage of this journey, he was always the last person to leave the bus or street car. After going through a long surveillance detection run, he changed into the Russian clothing that had been selected for him by CIA personnel foraging through Polish flea markets. At his final stop, he left the bus two miles from the manhole that was his target. Prior to this, he had never visited the site but had only seen photographs. He carefully checked for surveillance as he approached it. He used his OTS designed pry bar to lift the cover and when in the chamber he placed the sensor on the cable. In returning to his family picnic, he took a more direct route because surveillance was less important once his mission was completed.

CKTAW worked until 1985 when the officer who was supposed to recover the recordings received a "tamper indicated" signal and aborted his recovery mission. When another trip was made to the site, the tapes were recovered but the system had stopped working. Initially, there was speculation this might have been a technical problem. However, in August, 1985 a KGB defector informed the CIA that the CKTAW operation had been exposed by CIA employ who, after being fired, sold this information to the KGB.

Operations like CKTAW employ what is known as "large technology" as opposed to the useful gadgets used by operatives on the ground. Large technology is extremely expensive and requires a major effort in production. The KH-11 Kennan is an example of large technology. When the CIA needed equipment to be used for on-the-ground surveillance and counter surveillance it relied on its engineers in TSS and subsequent versions of TSS. There were two categories to be considered. The first was equipment used for offensive purposes such as getting information about recruitment targets. The second was for defensive purposes when counter surveillance was used to protect operatives involved in covert activities. If a case officer was going to meet an asset in a hostile region, counter surveillance was required to determine if the case officer was being watched as he went to the meeting. If long-term observations are being made from a building or some other fixed site, the operatives could use still cameras and video cameras to monitor the site. As a result of technological advances, it became possible to establish unmanned observation posts, so fewer personnel were required. Stationary surveillance required imaginative concealments such as the cuckoo clock used for hiding a camera to keep a room under observation. Mobile surveillance operatives have employed a camera hidden in a briefcase that would be positioned facing the target. A camera such as this did not have a viewfinder because the user simply held the briefcase under his arm. The East German Stasi developed a surveillance briefcase that could take infrared photographs in complete darkness. It used a camera with a silent electronic shutter. The case was covered by a special fabric so infrared light could reach inside the case. The cover shielded the infrared flashes so they could not be seen.

By the 1990s, surveillance photography had been greatly improved by miniaturization and greater camera storage capacity. Video cameras became more useful because they were much smaller and could operate in lower lighting conditions. As a result, covert collection of information was easier and more effective.

If a surveillant was fortunate enough to be close to the target, concealment of a traditional camera was possible simply by hiding it under clothing. Such devices as large belt buckles were useful and many cameras were designed so the lens could be hidden behind coat buttons. When pictures had to be taken at night, ultra-high-speed film made it possible to get a good picture with no more than the light from one candle. If you could change the developing time or temperature, the film would be even more effective. By 2001, digital imaging was increasingly common and effective. The Nikon DOX high resolution camera emerged as one of the most popular devices for photographic surveillance.

When a surveillance team did not enjoy the advantage of proximity to its target, there were cameras such as the Questar Seven 2,800mm. The effective range of this long lens innovation allowed it to read license plate numbers from two miles away.

Another valuable innovation for surveillance operations has been the development of sophisticated tracking devices or beacons. These may be as small as a credit card or, like the "parent beacon," as large as a WiFi router. The smallest beacons can be used for the tracking of assets that are being moved. If an adversary is traveling over great distances and is carrying, for example, a case full of money, the case can be tracked by a tiny beacon placed in its outer shell. There have been numerous cases in which terrorist equipment has been tracked while passing through several countries. Rather that have a surveillance team attempt to follow crates full of sensitive equipment, a well-placed beacon enables surveillance via a computer that traces a small chip. For especially complex operations in which surveillance is required on several targets, a "parent beacon" can be used to monitor multiple beacons. Again, this is done without endangering surveillance teams and no action needs to be taken until the shipments arrive at their destination in preparation to be used in an attack.

There are various types of beacons based on the purposes for which they are employed. For operations that are close to the target, a tactical beacon is used. One of the best-known tactical beacons is the PowerFlare PF-200. It is lightweight and uses a rugged infrared signal that produces a 360-degree infrared light. It can use several signal and flash patterns. It is currently being employed by security services, military, and law enforcement. Strategic surveillance uses beacons to monitor a broader range of activities that might constitute a threat. Its monitoring is based on selected information sources that draw from different disciplines. Satellites or aircraft are generally used to monitor strategic beacons. Most of these beacons—whether tactical or strategic—will employ a tiny radio frequency transmitter that serves as a navigational signal.

Because of their miniscule proportions, concealment is easy and the devices can be placed inside shipping containers when there is a need to track their movements. For people who are vulnerable to kidnapping, beacons can be placed inside shoes, belts, or some other article of clothing. Once that person is taken by kidnappers, he can activate the beacon which will transmit a signal. Another device that serves a similar function is a taggant, much like the security tag used by commercial establishments, which keeps track of the movements of the person being protected. The KGB was known for using a compound known as "spy dust." This mildly radioactive substance could be placed on clothing worn by a suspect or placed on the furniture in a room holding materials being sought by the suspect. If the "spy dust" could be detected on a suspected spy, it was proof he had entered the room.

In the decades after World War II, success in technical collections was dependent on audio operations and satellite photography. Audio operations assumed a high priority for the CIA engineers. The enemy communications system was the objective for audio operations. The effort put into the Berlin Tunnel operation was a tribute to the importance attached to our ability to listen in on the adversary's discussions. In that period, conventional landline telephones were especially vulnerable. When the British technicians penetrated the cable of the

Soviet headquarters, there was a flood of information that immediately became available to British and American intelligence services. A very simple innovation was created by the CIA's TSS when it devised a system that by-passed the telephone hook switch so the phone served as a microphone for listening to conversations in the room where it was connected.

Contemporary cell phones are even more vulnerable to audio surveillance. The information passing through a cell phone can effectively be plucked out of the ether. No physical connection is required. In fact, with simple technical adjustments, the cell phone is transformed into a spy that can transmit conversations, data, and the physical location of the user.

The most basic component of an audio operation is the contact microphone that is fixed to the wall of the target location. Sound causes vibrations on most hard surfaces and the contact microphone will pick up sound even through a concrete wall. Pinhole microphones will work with a tiny airway of less than half a millimeter in width and are more difficult to detect than the contact microphone. Even more sophisticated is the fiber optic microphone that uses cable that is smaller than a human hair. It uses light waves that can be transmitted on its miniscule cable placed under a door or through a hole in the wall. This remarkable device began as something used for medical procedures.

One difficulty with audio operations is that background noise often makes it hard to understand what your target is saying. The creation of the directional microphone—sometimes called acoustic radar—focuses on specific individuals rather than picking up a sea of incomprehensible noise and will focus on sound that is in front of the listener. It is intended for use in collecting human voice audio information. The directional microphone will pick up sound from a certain direction rather than from an entire location. These microphones may be omnidirectional (meaning a circular pickup pattern), unidirectional, or bidirectional. The basic part of the microphone is a waveguide pipe that has a diameter of 10mm to 30mm and individual cells that amplify sound. If listening to targets that are outside, there is a type of directional microphone known as a rifle or shotgun microphone that is effective outside. When a news crew attempts to interview a subject outside the building, it will use the rifle microphone. It will pick up sound from the direction of the target but filter out any noise from the side. This microphone is useful if your target happens to be a smoker forced to go outside for his smoking break.

When preparing an operational plan that involves audio or video recording, there are three basic components. The first is a recording device that will capture the information so it can be transmitted to the surveillance team base. Therefore, the second component is an effective transmission link. Finally, there must be a listening post that is reasonably close by the transmission that can be done by a wire rather than a less secure radio link.

If a team intends to use wire for transmitting the information, it can use several tools created by the CIA's Office of Technical Service (OTS). The best known is called the "fine wire kit" and includes tools that will help create separations for holding a wire. It offers a small crow bar for pulling back a baseboard so the tiny wire can be placed behind it. If there is no baseboard, the surveillance team can use the razor contained in the kit and create a small slit in the wall as a location for the wire.

It is important to be able to hide or disguise listening devices. Chinese companies routinely produce and sell a variety of tiny cameras that are designed to look like normal writing pens. These devices will also function as an actual writing implement. Many of these spy tools are voice activated and can be used for audio and video recording. The cheapest of them comes with a web camera and sells for around $40. The Chinese company even offers a pinhole camera, a motion detector, and advice on where to hide their cameras. During the Cold War, the Czechoslovakian security service developed a device known as the David pen. When the user removed the tip of this fully functional pen, it exposed a camera designed to take photographs of documents.

The OTS created a variety of gadgets that could be used by operatives who only had two or three minutes to plant their listening device. In an age when smoking was common, agents could carry what looked like a cigarette lighter but was actually a listening device. This might be casually left behind to pick up conversation in a room the agent had just left. A narcissistic person might use this to learn what people said about him after he left the room, but an intelligence operative could hear what plans might be developed by an enemy to counter their operations.

OTS also created an AC electrical adaptor that was actually a bugging device that could be placed between a lamp and the electrical outlet. It built furniture that hide concealment cavities for placement of listening devices or even blocks that might be placed inside existing furniture to hid a listening device. A Soviet musician named Leon Theremin created a passive cavity resonator that functioned like a drum. When targeted by an external transmitter using radio signals at the correct frequency in a building across the street, the resonator was activated to listen to conversations in a particular room. This is referred to as "illuminating" a passive device. The Soviets' passive cavity resonator was hidden inside a carved wooden version of the US "eagle" emblem that was presented to the US ambassador to the Soviet Union. Its true purpose was not exposed until a decade later and Ambassador Henry Cabot Lodge Jr. presented it to the United National General Assembly in an effort to embarrass the Soviet Union.

The CIA engineers created an assortment of concealment devices over the years. These included a lighter, a pen, and a key fob that contained the tiny Tropel cameras used in surveillance operations. One of these devices was used by Lieutenant Colonel Boris Yuzhin, who was a mole inside the KGB and was actually spying for the FBI. While visiting the Soviet Consulate in San Francisco, he left his Tropel camera concealment behind, an action which prompted the KGB to begin an investigation that led to his eventual imprisonment.

The KGB designed and used a fixed surveillance camera that could be easily hidden inside common items. Their most successful employment was inside a decorative mask given as a gift to US officials. The tiny camera was set to take pictures at certain intervals.

Modern technology represented a threat to American interests not only with the placement of the Theremin device in the office of the US ambassador to the Soviet Union. Through the NSA GUNMAN project, the NSA uncovered evidence of another unique and sophisticated Soviet effort to penetrate the US Embassy in Moscow. In 1976, the KGB managed to install miniaturized electronic eavesdropping equipment and burst transmitters inside sixteen IBM Selectric typewriters used in American diplomatic facilities in Moscow and Leningrad. This equipment copied everything typed on those machines and then transmitted that information to nearby KGB listening posts. This system operated for eight years and was not discovered until President Reagan authorized the GUNMAN project in 1984. Within one hundred days, an NSA team replaced every piece of communications and encryption equipment and all of the computers, typewriters, and printers used in Moscow and Leningrad. Only then did they uncover a sophisticated and destructive Soviet operation.

More recently, thanks to the proliferation of cheap surveillance technology, almost any government or individual can become a surveillance power. A Virginia based satellite operator known as HawkEye 360 offers the services of its CubeSats which can be used with almost any type of equipment. The CubeSats are nanosatellites that are no bigger than a shoebox. Its standard size measures 10x10x10 centimeters, although there are larger ones that are the size of a school bus. They were developed in 1999 and designed to be a platform for education and space exploration. Over the past two decades, the CubeSats have become an important part of industrial and governmental markets. The first CubeSat used as a spacecraft was launched in 2006 by NASA Ames. Since then it has launched sixteen more CubeSats of varying sizes.

The most successful marketer of the CubeSats is probably HawkEye 360. It is a Radio Frequency geospatial analytics company based in Herndon, Virginia. It operates a commercial satellite constellation that can identify and locate RF signals. Their services recently assisted Ecuadorian authorities faced with the problem of illegal fishing being conducted by Chinese vessels. This problem is not limited to just Ecuador. There is a multitude of gangs involved in smuggling, piracy, and human trafficking around the world. The countering of these threats requires effective monitoring of vessels that routinely deactivate the Automatic Identification Systems that are supposed to monitor their movements. When this happens, nations can only defend their interests by dispatching coast guard vessels or airplanes to search the sea. These are extremely expensive and beyond the financial reach of many nations. HawkEye 360, with its fleet of small commercial satellites, can expand the visibility of such nations so they can locate the "dark ships" that are involved in criminal behavior. HawkEye can pick up those ships' radio frequency signals that are used for marine radar and satellite communications.

HawkEye360 uses trilateration or triangulation to locate ships that have turned off their Automatic Identification Systems. As an alternative, it can measure small shifts in the signal frequency of the ship's transmitter as it moves. Given the commercial success of HawkEye360, it is not surprising that other companies are being set up to benefit from the need for more effective surveillance at sea. HawkEye's data can effectively locate guerrilla camps and mobile missile-launchers used by terrorist organizations.

With its reliance on satellite clusters to collect data based on RF signals, there is speculation that the technology of HawkEye360 may soon fall behind the times. Engineers at the French-based Unseenlabs have determined that the clustering technique is no longer the best for use in monitoring ships at sea. The detection system used by Unseenlabs requires only a single satellite instead of the cluster presumably required for accurate triangulation. How their system works is a vital and well-protected secret that is guarded by French security services. If Unseenlabs enjoys a monopoly on the single-satellite-RF-intelligence market, it is unlikely they will enjoy this advantage for much longer. Its most significant emerging competition comes from Horizon Satellite, an Alaska-based company founded in 1990. According to Horizon engineers, a single satellite can determine the location of its target by using differences in the angles of a target's signals as it travels across the sky. They maintain that this gives them the ability to pinpoint the target's location within 3,000 meters while Unseenlabs can only guarantee a location within 5,000 meters. They claim their technology will also enable Horizon to create a library of radar-pulse "fingerprints" so any vessel can be easily identified no matter where it may be located.

Another blessing of the new technologies is that they have made it possible to spy on internet users who attempt to maintain anonymity. The Tor network is one of the most frequently used browsers by those hoping to hide their identities. Working together, the NSA and Britain's GCHQ have sometimes been able to block access to the TOR network or to actually expose the identity of TOR users. By implanting malicious code into the computers operated by TOR network users, clandestine operations have disabled elements of the anonymous community. They have also targeted members of the internet group known as "Anonymous" which has been responsible for cyber-attacks against major corporations, governmental institutions and agencies.

The same technology enabled Britain's GCHQ to set up an automated system to monitor reservation systems in certain luxury hotels. There were 350 hotels that were targeted by the GCHQ, all of them luxury hotels around the world. Known as the Royal Concierge surveillance program, it reflected their interest in monitoring the activities of numerous random individuals determined to be of possible natural security interest. They could also wiretap telephone calls made from the rooms in these hotels.

Working together, NSA ████████ and GCHQ even conducted surveillance on many online games. The specific targets were the massive multiplayer games that involved role-playing. The objectives of this virtual reality surveillance are not immediately apparent, but it may be a reflection of the desire to do something simply because it is possible. The value of this endeavor may be a function of the social interactions that occur both within and outside of such games.

Political espionage is an obvious aspect of surveillance operations. Intelligence organizations typically establish specific intelligence priorities. For the NSA, this means setting a scale of one to five with one representing the most important and five indicating the lowest level of interest. The NSA collaborates with thirty nations in its surveillance endeavors, although it also spies on many of these partners.

██

████████████████ The ██████████ Special
Collection Service ███████████████████████

██

███████████████████ was established in 1978 at the height
of the Cold War. ████████

██

SCS operatives are believed to have hidden listening devices under the wings of pigeons trained to perch on the window sills of the Soviet Embassy in Washington. ████████████

██

Throughout the global war on terror, the SCS was instrumental in establishing eavesdropping post in major cities in the Middle East. It was able to monitor al-Qaeda training camps and successfully target known associates of Osama bin Laden. When bin Laden's compound in Pakistan was located, SCS operatives set up a base in an apartment one mile away and used lasers to target the compound windows. By analyzing vibrations, they determined how many people were in the facility and that one of those individuals never went outside. They correctly surmised that this was Osama bin Laden, thus making possible the special operations raid in which bin Laden was killed.

CHAPTER 14

Privatization

A NOTHER EMERGING PROBLEM ASSOCIATED WITH THE remarkable growth and sophistication of new technologies is the development of private intelligence companies. These technologies were initially available only to government agencies with generous budgets. Now they are cheaper and commercially available to private entities. Because of constitutional limits on how the CIA or the FBI might take advantage of the new technologies, in 2021 the Department of Homeland Security began considering the employment of private entities to conduct surveillance on citizens who organize resistance to federal policies. Unlike federal agencies, private companies can assume false identities to gain entry into private messaging apps used by dissident groups. In addition, private contractors do not need warrants to gather this type of information.

Before the emergence of private companies doing the work of the intelligence community, private contractors were being recruited to handle specific tasks. For the US intelligence community, the contractor route was effective and contributed to the development of intelligence tools during World War II. Private firms were increasingly recruited because they had the skills and facilities needed to build the arsenal of weapons needed for the Office of Strategic Services (the CIA's predecessor organization) and other such agencies. Through its Office of Scientific Research and Development, the OSS was able to identify and enlist contractors who could quickly produce highly specialized military related items on demand. These innovators popularized the idea that technology could win the war.

By contrast, the British system involved purchasing the laboratories needed for this work. The British countryside was cluttered with large, splendid estates owned by old families whose resources had degenerated during and after World War I. Consequently, they could no longer maintain those estates and were happy to sell them to the government, which turned them into research institutes for the development of weapons with which to conduct special operations. Although secret at the time, Aston House today is recognized as the location where the first time fuses were developed. Aston House was joined by The Frythe where special weapons were packaged and sent out to OSS operatives. Soon, the British government owned and operated estates that became vital to the wartime effort.

The private American firms that worked over the decades fulfilling often small specialized orders did not get rich from their business with the intelligence community. They did, however, create innovations that emerged as popular consumer items that shape contemporary American life. Through research for the Intelligence Community, companies found themselves being able to market everyday items such as the GPS, text messaging systems, and other commercial successes. Intelligence Community relationships with private firms have been driven by a need for the technical capabilities enjoyed by the private sector. By working with a private aeronautics firm, the CIA was able to develop the U-2 spy plane. This ambitious project would never have been put into operation without the services of private industry. The Air Force would never have had the MQ-1 Predator drone without its private sector partnership.

Employees of private firms often have skills needed by the Intelligence Community. Those skills range from security services for essential installations and language skills to information technology skills needed to maintain interagency coordination. Moreover, private firms offer a degree of flexibility absent within the government work force. If the Intelligence Community recognizes a sudden, unanticipated requirement, it is convenient to enlist teams from a private firm. Since those workers will not enjoy the civil service protection available for regular government employees, they can be dismissed once the urgent need has passed. The Intelligence Community manpower shortage may well be the result of a lack of long-term planning but this option protects it from the consequences of poor planning. The contractors can also fulfill mundane tasks such as data entry that might not be appropriate for more senior government employees.

It is difficult to measure the full extent of the corporate presence within the intelligence community. Press reports have indicated that 51 percent of the workforce of the Defense Intelligence Agency is composed of contractors. A 2007 report by the Director of National Intelligence suggested that contracts with private firms account for 71 percent of the Intelligence Community budget. A Senate Intelligence committee report found that the services of a contractor cost twice as much as a Civil Service employee. If that is true, it would explain the disparity in the Intelligence Community's budget.

The privatization of intelligence began with modest steps, such as contracts to develop a specific technical device or the recruitment of individuals to do a one-time task. From that point, the contractors became increasingly well organized and surprisingly well known. In the United States, Booz Allen Hamilton, Fusion GPS, and Stratfor have emerged as among the most prominent. In Britain, Control Risks Group and Cambridge Analytica were two of the best known. Cambridge Analytica, after illegally obtaining personal data of millions of Facebook users, was embroiled in controversy and forced to close.

The recent mergers of already powerful intelligence companies have created giant entities with frightening monopolist tendencies. There are now five companies that, together, employ 80 percent of the private sector contractors working for surveillance and spy agencies. Leidos Holdings, after completing a merger with Lockheed Martin's Information Systems & Global Solutions division, may be the most powerful company in the world of intelligence-contracting. Other companies in the top tier of intelligence contractors are Booz Allen Hamilton, CSRA, SAIC, and CACI International. The monopolization of intelligence contractors began in the late 1990s and fundamentally changed what was previously a diverse industry.

Because these five companies employ 80 percent of the private sector contractors, if a contract entity makes a serious mistake, the government has a limited number of options. Moreover, so much of their work is top secret, there is very little reporting about their activities, a fact that undermines their accountability. When a contractor working overseas engages in questionable practices, it is difficult to determine who has the authority to deal with the resultant problems.

When thinking about private intelligence services, the most common assumption is that government is reaching out to the private sector. There have been situations in which government actors are irrelevant. Histories of the Coca Cola Company often observe that the company has one of the oldest and most effective corporate intelligence divisions. It is obvious that if your business operates around the world, you need to know how local conditions might affect your operations. You will also want to know what your competition is doing so corporate espionage is important.

In previous decades, such companies ran their own intelligence divisions. However, just like the government, their operational model had been changed by the advancement in technology. Therefore, like government agencies, they have turned to private contractors. Not only do the specialized contractors have skills and experiences beyond what your company can do, if they get caught doing something illegal, the company itself has plausible deniability. If a company needs technical surveillance countermeasures, they can turn to a firm like Deutsche Telekom which offers its many services to a variety of customers. They can routinely perform operations such as bug sweeping services to protect a company that feels threatened by its competitors.

There are, however, limits to what a private firm can offer. Many of these firms are staffed by "consultants" who often produce analysis based primarily on open source materials. In the world of the internet, the scope of open source materials should never be discounted. Within the world of governmental intelligence services, there is often a tendency to dismiss open source materials. Any examination of the CIA product once known as the Foreign Broadcast Information Service will disabuse you of the notion that the CIA ignores open sources. In the offices of most analysts, you will see copies of this service on desks or in trash cans. The private consultants may be reading the same reports but they can develop useful studies of important issues. In the end, it is clear that few private contractors have developed independent collection capabilities. First, there are legal barriers. The private contractors cannot get a warrant to tap into a company's telephone network, for example. Second and probably even more important, the cultivation of private sources is an extremely expensive process and the financial rewards are not sufficient to justify this.

Intelligence work such as surveillance is heavily dependent upon private actors. Companies such as Booz Allen Hamilton, Google and Facebook have become a major part of the world of non-governmental agencies that can perform tasks for the government. Intelligence-related government agencies depend on Amazon Web Services for their cloud computing infrastructure. Other companies help analyze data needed by the CIA, NSA, the FBI, and a variety of local government agencies. It is, however, important to remember that while they can assist the government, their greatest function is to serve the bottom line on their annual earnings statement.

CHAPTER 15

Evolution of Surveillance

THERE IS NO WAY TO MINIMIZE the importance of technology as we examine the evolution of surveillance. The new technology has made possible tasks that would have previously have been unimaginable. In a broader sense it is more important to consider the changing role of surveillance over the decades. It affects our lives far more dramatically than at any time in the past. Within the context of our lives, surveillance has become more prevalent and determinative of the evolution of our society itself. Because of this, there are debates about "data mining" and the extent to which this might undermine a free society.

Data mining, otherwise known as knowledge discovery in data, has become very important in recent years and has become controversial because of its use in predicting behavior. The goal of the process is to extract patterns of knowledge that can be determined from tremendous storehouses of information. With the data warehousing capabilities being used by big data, data mining techniques are being used to enhance the profitability of corporations. Criticisms have been based on the predictive capabilities made possible by the use of machine learning algorithms. Typical objectives include the detection of fraud and the exposure of problems with security procedures.

Objectives associated with law enforcement are even more problematic as data mining has become a predictive policing tool. One such tool known as PredPol is being used by dozens of cities in the United States. PredPol divides communities into 500-by-500-foot blocks and makes predictions about future criminal behavior throughout each day. Using data about people, it predicts who is most likely to be involved in criminal activities each day. This information could make it possible for the police to intervene before a crime takes place or a judge could use the data to predict who is likely to be a repeat offender and render a sentence based on that assumption. The predictive algorithms used in such programs are often based on arrest records. In spite of the shaky foundations of these programs, their defenders argue that an algorithm is less likely to display the bias often associated with human behavior. By contrast, critics have complained that this is simply another instance of data being weaponized against minority communities and is, therefore, racist.

Within the intelligence community, officials tend to defend their collection projects against the accusation they are engaging in data mining. The willingness of Facebook to sell information about 50 million clients contributed to the controversies around this process. Cambridge Analytica appeared to enjoy enormous profit from this activity. Because of scandals such as this, data mining seems to be irrevocably associated with the idea of selling private user information in order to make money. There is also suspicion because of claims to be able to predict outcomes relative to business or even political trends. In spite of controversies, any examination of employment websites will produce numerous hits for data mining and other related job openings.

Technology, of course, has revolutionized almost every aspect of our lives. It has driven the evolution of surveillance techniques away from the simple matter of having an operative watch the movements of a target moving through a city. The lone operative has been increasingly displaced by satellite systems such as Corona photo-reconnaissance system, which demonstrated its first success in 1960.

In 1976, the KH-11 system produced "real-time digital imagery" that enabled surveillance operations to rise to an entirely new level of precision. The Defense Advanced Research Projects Agency (DARPA) emerged as the CIA's most reliable partner in creating a new generation of surveillance systems that enhanced capabilities while reducing the risks to individual operatives. In the not too distant past surveillance operations were in the hands of a small group of operatives who risked imprisonment by brutal regimes determined to hide their secrets. Now the skills of those small teams are augmented technological innovations that would have been unimaginable in the early days of the Cold War.

While these large technology advances are invaluable to surveillance operations conducted by governments, they are well beyond the reach of most civilian surveillance companies. However, there are numerous small and affordable gadgets that can aid the private detective who may be running surveillance on a wayward spouse. It is equally important that many of these devices are not beyond the budgets of many companies. The proliferation of private intelligence organizations has been fueled, in part, by the fact that they can augment governmental services and do not labor under legal restrictions that reign in government agencies.

The latter feature of contracting work raises questions about the possible evolution of surveillance related activities. Who is responsible for the oversight and management of the work of private contractors? It is obvious that Congress will have to assume responsibility for this matter. There will need to be clear guidelines to determine when it is acceptable for contractors to undertake sensitive intelligence tasks. In the event of war, these questions become even more compelling and urgent. The outsourcing of government functions may seem attractive to many people but the process threatens to undermine constitutional limits on government.

As our national leadership has recognized the growing power of the tools used in intelligence and surveillance, there has been an effort to maintain a delicate balance between citizens' freedoms and the need for national security. Before technology was an important factor in strengthening the nation's police powers, America's Founding Fathers endeavored to protect citizens against unreasonable searches and seizures. In 1791, this was expressed in the Fourth Amendment to the Constitution. It was not until 1928 that the appearance of new technology moved this concern into the legal spotlight. In the US Supreme Court case Olmstead v. United States, there was a challenge to the practice whereby federal officials collected evidence from wiretaps placed without judicial approval. The ability to use wiretaps of telephonic conversations was a result of a relatively new technological innovation and the Supreme Court approved the practice. A phone call, after all, was not a physical artifact so it did not enjoy Fourth Amendment protection.

Wiretapping is another area in which private companies led the way. Going back to the 1920s, it was private detectives and corporations that used wiretapping to gather information. Corporations employed wiretapping as a way of suppressing unionization efforts. The increased demands during the Prohibition era forced law enforcement to rely on wiretapping.

Over the years, both public and official attitudes toward wiretapping were relatively relaxed. It was with some reluctance that wiretapping for national security reasons was finally accepted. However, wiretapping to catch tax evaders was a different and more troubling matter.

Continuing along these relatively permissive lines regarding surveillance, in 1934 the Federal Communications Act addressed wiretapping and determined that it was legal to collect information in this way as long as that information was not shared outside law enforcement institutions. In 1945, project SHAMROCK was created by the Armed Forces Security Agency. The purpose of this project was to collect information drawn from international telegrams without the requirement of a warrant. In 1952, the work of the Armed Forces Security Agency was taken over by the National Security Agency and SHAMROCK's surveillance of international telegram traffic continued.

The government's permissive policies were finally challenged in 1967 when the Supreme Court overturned the Olmstead precedent in Katz v. United States. In the Katz decision, the Court ruled that non-tangible possessions—phone calls and electronic transmissions—were protected by the Fourth Amendment. In 1968, Congress enacted the Omnibus Crime Control and Safe Streets Act that restricted wiretapping. Only the President, in service to national security interests, would be allowed to make an exception regarding federal wiretapping.

Further scrutiny came about as a result of the 1972 Watergate scandal and the hearing of the House Judiciary Committee, which issued articles of impeachment against President Richard Nixon based on illegal wiretapping, misuse of the CIA, and other abuses. In 1975, the Church Committee investigated activities of the CIA and the FBI and revealed that there had been hundreds of cases of warrantless wiretappings and illegal electronic surveillance.

In 1978 Congress passed the Foreign Intelligence Surveillance Act (FISA). Under this act, there would be a secret court that would consider requests for warrants to conduct electronic surveillance. Protection for cell phone conversations and internet communication were added to the Omnibus Crime Control and Safe Streets Act in 1986.

As a result of the terrorist attacks of September 11, 2001 and the passage of the USA Patriot Act, there were changes to FISA procedures and wiretap restrictions became more flexible. In 2003, an AT&T technician discovered a secret room being used by the NSA to conduct data mining operations of internet traffic. In 2005, there were further reports about extensive use of government surveillance to conduct warrantless wiretapping and surveillance operations against internet users. It is generally acknowledged that the US government has complete access to any data that US-based phone companies have collected.

Contemporary electronic surveillance is dramatically different from what took place in earlier times. Until the 1980s, it was largely individualized and focused on specific personalities. The surveillance operations themselves were more specific as law enforcement was investigating known or suspected criminal activities. More recent electronic surveillance is broadly based and not targeted on certain suspects. It is often referred to as dataveillance and is less concerned with what is said in a conversation. Now the surveillance operation is looking for information about who a person called, when the call was made, and where the phone was located. The target is more likely to be the metadata of particular financial transactions.

As can be seen by more recent efforts to limit wiretapping and surveillance by government agencies, there is now a great divide between government surveillance and the activities of private intelligence companies. The current debate is focused on limiting or managing governmental surveillance programs. Private operations are more likely to be under the radar and not subject to official studies or limitations.

There are also legitimate questions about the morality of surveillance. Most citizens feel suspicious of the growing ability of the state to monitor almost any aspect of our public or private behavior. At some point it is essential to weigh the benefits of surveillance against the possible negative impact of surveillance. Even people who are not guilty of illegal behavior may be subject to surveillance and the resultant records may remain on file indefinitely. An average citizen may be justifiably fearful that those records might be used against him at some future date. This knowledge will inevitably have a chilling effect on citizens who have no intention of breaking the law. This fear may well produce a feeling that we need to conform and we must repress the creativity that might make one appear different. The result will be a popular reluctance to express opinions or to become active in the governance of your community.

It is true that the operational capabilities for large surveillance activities are beyond the budgets of most companies advertising their ability to manage surveillance over physical facilities. Their collection activities are relatively mundane, even when they hold federal contracts to supplement governmental collection efforts.

As individual entrepreneurs marketing their services, they are limited. However, the greatest power of private sector capabilities is in the area of management of data, a capability that allows them to overcome existing limits. The internet, from e-commerce to Google, has given private entities the ability to accumulate more information than is held by not only the US government but also probably by all governments combined. As they collect information about citizens, they are not limited by the Fourth Amendment of the US Constitution. By offering services and products, the internet enlists people to voluntarily reveal an enormous amount of personal information.

So, while internet giants like Google face fewer restrictions in conducting surveillance of their users, they are still obligated to share that information with US governmental agencies if they are presented with relevant legal claims. Under certain circumstances they can be required to reveal the actual content of private emails. What we see here is that the power of the private sector complements the authority of the state in order to have full surveillance of the activities of citizens. In addition, many governments have entered into partnerships for sharing information they have collected. The first nations to do this were the United States, the United Kingdom, Canada, New Zealand, and Australia. Since creation of this so-called "Five Eyes Alliance," several other governments have been included in this arrangement.

Citizens who hope to ensure their privacy can utilize a VPN service or they can shift their location in order to get a new IP address. But the government has unlimited resources and can defeat most efforts to protect information. Moreover, most governments have passed or hope to pass legislation to require internet service providers to retain all data about the internet activities of their customers. Some governments are trying to establish a legal framework under which they could utilize Trojan Horse software so that agencies would always have a window into civilian databases.

For years debates about the legitimacy of surveillance have been shaped by the notion of a reasonable expectation of privacy. There was a comfortable assumption that many things should be entirely beyond the range of surveillance. Therefore, what a person did on the street was very different from what might be done in the home. There was no reasonable expectation of privacy regarding activities on the street. As more and more of our activities were conducted on the internet, the notion of reasonable expectation of privacy was under pressure. Online surveillance is routinely undertaken by the state. At the same time, state cyber-policing of the internet in an effort to prevent certain types of crimes is also accepted policy. The authority of regulatory agencies has expanded to cover many internet-based activities and, as a result, there is much less of an expectation of privacy when using the internet. Consequently, people are reluctant to exercise constitutionally protected freedoms or even to engage in legitimate activities on the internet. When we assume that the state is watching, it is not so different from how a driver feels when being stopped by police while driving. There is a fear that you may have unknowingly violated an obscure law. With the abundance of laws regarding what is legal or illegal on the internet, it often seems safer to do nothing. Surveillance, by itself, can deter people from exercising legitimate rights. This phenomenon has sparked a debate about the chilling effects of regulatory actions that are becoming routine.

One of the great challenges that exist today is that our scientists move rapidly in developing surveillance technologies needed by both government and the private sector. At the same time, the speed of the scientists cannot be matched by the lawmakers who are supposed to maintain and develop a legal framework to protect the privacy of citizens. Legislative bodies have not been able to alleviate the disparity of power that is a result of the massive collection capabilities that threaten to overwhelm the freedoms of citizens. The combined resources of the state and the powerful corporations working with it have become an insurmountable Leviathan that is beyond the reach of a countervailing power.

It has long been said that knowledge is power. The development of modern technology has created devices that can collect infinite amounts of information. As the state, which already has a monopoly of force, developed its hold on mountains of data, its authority grew beyond the ability of individuals to limit its control over society. Corporate entities, once primarily based within the nation, have expanded their reach to global dimensions. Thus, multi-national corporations now have virtually unlimited control over citizens. The instrument for utilization of this power is now social media which can shape public debates and influence the outcome of elections. As a result, social media, operating as a utility, determines who gets jobs, how contracts are awarded, and what constitutes an acceptable component of our culture.

What is now referred to as "big tech" has a degree of power that would have been unimaginable a few decades ago. The most miniscule information about a person can be marshalled to paint an unrecognizable image. In the not too distant past, it was possible for people's past misdeeds to be forgotten or even forgiven. That is no longer possible and something as trivial as misbehavior in high school can continue to appear in records that may become public if you are about to receive some prestigious honor such as a nomination to the US Supreme Court. By mounting a concerted effort using the internet, no person is immune to fatal levels of character assassination.

A pseudo-scientific psychological profile can be shaped by presenting an assortment of facts in a destructive fashion. The result may well be an image that the individual would not recognize as himself. The specter of such an attack serves to prevent many citizens from entering politics or even participating in various community functions. Because social media platforms were designed to be addictive, millions of people are drawn to them and their opinions or even self-images are shaped by what appears on those platforms. Through their manipulation of massive amounts of data, the big tech companies have amassed unprecedented power that rivals the actual power of the state in many respects.

As noted above, when the government decides to outsource surveillance to private companies, it is able to defeat requirements for transparency. Those private companies are not accountable to the public and cannot be subjected to a FOIA request or any other form of protective legal device. Perceived threats to the public—terrorism, for example—are used as justification of draconian measures that would have been inconceivable even during the Cold War. Because there are so many opportunities for enrichment, these private companies have a constant motive to track their fellow citizens.

Companies tout the economic benefits of modern surveillance technology. Canadian insurance providers argue that greater use of these technologies will have an immediate benefit to consumers who will experience lower insurance costs as fraud can be more easily detected. Most importantly, insurance providers say it will be possible to reward "good" customers who always comply with safety requirements. At the same time, customers exhibiting "reckless" behavior will be punished with higher rates.

There is a widespread willingness to accept restrictions when people are motivated by fear. Fear of health threats enabled officials to assume emergency powers in response to the COVID-19 pandemic. Much of this work is being undertaken by TECH5, a Geneva-based international technology company that is developing biometric-driven identity management solutions. It has taken the lead in a health surveillance program described as a digital health certificate. Seven EU nations—Bulgaria, Croatia, the Czech Republic, Denmark, Germany, Greece, and Poland—were early participants in this program. Holders of the digital health certificates will be able to travel in any of the participating member states.

COVID-19 is also helping transform biometric surveillance into a routine technological application to promote health maintenance. BioButtons, devices that monitor an individual's temperature, heart rate, respiration, and location, were introduced at Michigan's Oakland University as a requirement for all returning students. In Ontario, the provincial government introduced a plan under which students would wear a contact tracing wristband that would vibrate when the wearer gets within six feet of a COVID-positive person. The data would be preserved in order to determine patterns of association.

Of even more immediate value, according to surveillance advocates, is the protection of citizens' property. Among the first to see this advantage was Amazon, a company whose survival is dependent on their ability to leave an unattended package at your door. Amazon could sell security to customers willing to purchase the Ring app, a camera-equipped doorbell designed to catch package thieves.

The Ring can protect your front porch, but another company guarantees to transform your entire neighborhood into a digital gated community. Flock Safety is a new technology firm that sells cameras directly to civilians. Police departments have long been using automatic license-plate readers that are sold by technology vendors. By contrast, Flock Safety markets their products to private citizens whom it can approach through homeowners' associations. In its advertising, Flock Safety maintains that its cameras are used in thirty-five states and claims that law enforcement personnel cite Flock's role in reducing crime. When Flock's cameras operate in a neighborhood, it records everything that happens and checks every automobile entering the area to determine if it belongs to a resident. The company promises that facial recognition programs will soon be added to their cameras.

Conclusion

THE EVOLUTION OF SURVEILLANCE IS NOT limited to issues of new technologies and how they may be used. There is a qualitative difference between monitoring the movements of a single individual or a small group at specific times and maintaining constant community wide or nationwide surveillance.

There is also an important question about relationship surveillance and oppression. An oppressive society requires a system for watching what people are doing and, if necessary, taking action against people whose behavior does not reflect officially sanctioned values.

Surveillance operations such as those discussed above are, in effect, tactical and not aimed at enforcing a "social credit" system as seen in contemporary China. In fact, the Chinese system raises the question of whether or not a government can collect information for purely benign purposes. If the information is stored as a permanent record of the individual's behavior, can it be utilized at a future date to pose an indictment against that person? Even if it is not, the citizen's knowledge of this system is likely to affect his behavior and deter him from expressing negative opinions about official policies. Thus, it facilitates social engineering and behavioral conditioning without actually imprisoning its targets. Chinese people see the impact of state-run social engineering efforts and realize its impact when people with low "social credit scores" are unable to travel or enjoy other benefits.

CCTV systems are increasingly being introduced into school. In 2021, the Springfield, Massachusetts School Board announced that it had joined with local police in installing a CCTV system in their high school and that it would be coupled with devices to detect weapons taken into the school or the use of vaping in school toilets. This program is consistent with the many camera-share initiatives that have been created throughout much of the United States whereby the public can be directly involved in law enforcement.

Popular involvement in law enforcement is being enhanced by the introduction of high-tech policing into smaller cities and towns. Data-driven policing, once confined to big cities, is becoming available in smaller jurisdictions. The Forsyth County, North Carolina Sheriff's Department is like other small communities that have fewer resources but are attracted to images that will transform them into a "modern" environment of mass surveillance normally associated with places like New York City. When the Pasco County Florida Sheriff's Department adopted a data-driven predictive policing system designed for at-risk young people, it announced that the data would be drawn from risk factors such as students' bad grades and arrest records. Even though technologically enhanced policing is attractive because it sounds modern, it raises questions about privacy and surveillance.

More and more people look at this system of massive surveillance and are pleased rather than terrified. They see the "kind hand" of a system that helps you locate the products you want even before you realized you wanted them. A device like Alexa is seen as your friend and servant, although it is essentially a wiretap. It is simply the latest form of corporate surveillance. It is always observing your behavior but, a person concludes, it is for your own good.

APPENDIX I

Static and Mobile Surveillance

ESTABLISHING SUCCESSFUL SURVEILLANCE BEGINS WITH THE setup. There is no reason why surveillance has to risk exposure from the very beginning of an operation. Indeed, operatives can set up around a target far enough away to not risk being seen, yet still close enough to provide blanket coverage.

In Figure 1, the star indicates the location of the target. The location could be anything—an office building, a home, or an embassy, for example.

Figure 1.

Note that in Figure 2, the surveillants are widely dispersed. One has eyes on the target, but the other two are far enough away so as not to be seen—but are positioned in such a way so as to be on the subject not matter what direction he chooses to travel.

Figure 2.

Figure 3 is clearly a public park. The constant movement of traffic and people in and out of the area will provide cover for the surveillants at the outset. But also note that the surrounding area is residential. Setup is easy as the target enjoys a coffee at the outdoor café, but once he moves into the neighborhood, it will be more difficult to track his movements without being seen.

Figure 3.

It is for that reason that the surveillants are positioned at the four corners of the park. Only one at a time must run the risk of exposure, and the four can then pass the subject off to one another as he moves through the neighborhood. In a situation like this, proper communication between the four surveillants is a must. Static surveillance in the area also would be helpful in Figure 4 that the surveillants could allow the subject to think he is not being followed while they regroup out of sight. In fact, the static surveillant would be watching the target from a building.

Figure 4.

The area in Figure 5 poses a much more serious challenge to surveil-lants. The area is spread out, likely located in a desert, and industrial in nature. What reason would a surveillant have for just hanging around in such an area? Cover for action would be critical. Perhaps surveil-lants would be dressed in work uniforms. Perhaps they would be driving dusty pick-up trucks rather than Mercedes sedans. Everything about them would have to fit in with the area and with people in the area.

Figure 5.

One idea for setup might be to have surveillants far enough apart from each other that no one person would notice them working together. And besides the surveillants' locations in Figure 6, another car should be farther down the highway, ready to intercept the target once he starts driving. The remaining four surveillants could catch up later and could then begin to pass the subject off to one another.

Figure 6.

APPENDIX 2

Sample Surveillance Detection Route, Washington, DC

B ELOW IS A SAMPLE SURVEILLANCE DETECTION route. The route begins at the corner of 20th and F Streets, NW on the campus of George Washington University, four blocks directly west of the White House. GWU is outside the White House "red zone," which should be avoided because of the intense security in the area. Remember, Phase 1 of the SDR should be the one in which the person running the route is going about his or her normal business. If your first stop is a straight shot from your kick-off point, then by all means go straight so as not to pique the interest of possible surveillants.

Depart from the corner of 20th and F Streets (Figure 7) and go directly north on 20th Street toward Dupont Circle (Figures 8 and 9). Go around the circle to the west side of Connecticut Avenue, NW and go into the Kramerbooks and Afterwords bookstore (Figure 10). Kramerbooks is a very popular bookstore and café that also has a back exit that will allow you to depart via the less-trafficked 19th Street. Browse for books and/or have a coffee, then depart from the back exit, go north to the end of the block, and make a left on Q Street, NW (Figure 11). This will begin Phase 2 of the SDR, the pattern phase. On Q, cross Connecticut Avenue, then turn south (left) onto 21st Street. Go one block and turn west (right) on P Street.

P Street is an important segment because it is the only thing that connects Dupont Circle with the Georgetown Section of Washington (Figure 12). It's a chokepoint that will allow you to funnel surveillance. If somebody is following you, they will have to cross the P

Street Bridge along with you to come out in Georgetown on the other end.

Continue west on P Street until you get to 28th Street. There you can go into Stachowski's, a high-end specialty meat market and deli (Figure 13). Browse the special cuts of expensive meats, maybe buy a sandwich or a drink, and exit the store, heading north on 28th Street (Figure 14). Go north one block and then make turn west (left) on Q Street. Turn north (right) on 29th Street and follow it one block to R Street. Turn west (left) and go one block to the entrance of Oak Hill Cemetery (Figure 15.) This is where you will enter Phase 3, the aggressive phase.

Walk around the historic cemetery. Perhaps you can spend some time at the graves of former Secretary of State Dean Acheson; President Lincoln's personal secretary John Nicolay; former Treasury Secretary Salomon P. Chase, whose image appears on the $10,000 bill; or former Washington Post publisher Katherine Graham.

When exiting the cemetery, proceed south on 30th Street into the heart of Georgetown. Here you will use the "stairstepping" technique, which, as you can see on the map, makes a series of right and left turns at every block look like a set of stairs. Your goal is to end up at the corner of 36th Street and Prospect Streets, NW, the house where the movie *The Exorcist* was filmed (Figure 16).

Immediately to the left of the house are the famous "Exorcist Steps," which attract tourists 365 days a year. This is the provocative phase, where you will either abort your mission or continue down the steps to a waiting car parked at the abandoned Exxon station below (Figures 17 and 18). Your route is now complete.

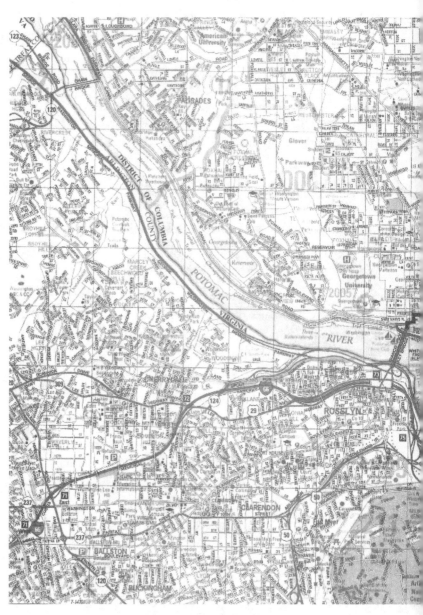

Washington, DC surveillance detection route.

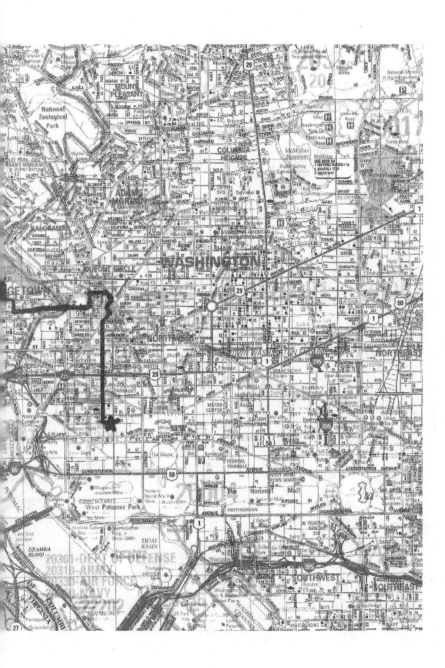

Figure 7.

Figure 8.

Figure 9.

Figure 10.

Figure 11.

Figure 12.

Figure 13.

Figure 14.

Figure 15.

Figure 16.

Figure 17.

Figure 18.

Sample Surveillance Detection Route, Alexandria, Virginia

SOMETIMES A SURVEILLANCE DETECTION ROUTE MUST be done by foot, not because it's necessarily easier that way, but because it is virtually impossible for you to be followed by car. Cut-throughs, one-way streets, and pedestrian paths become your best friends. And cemeteries are your friend: lots of people visit cemeteries, and surveillants who remain in their cars stick out like sore thumbs. Here is an example of a good foot SDR.

Using the Alexandria, VA courthouse as a kick-off point (Figure 19), travel east (left) on King Street toward the Potomac River (Figure 20). Walk three blocks and enter The Torpedo Factory, an artists' colony and headquarters of the Alexandria Archeological Society (Figure 21). Browse the artworks for sale on three levels and walk through the Alexandria Archeological Museum. Depart the building from the back exit along the river (Figure 22) and continue north (left) along the building.

Make a left at the end of the building and then turn north (right) onto Union Street (Figure 23). Continue on Union Street five blocks (Figure 24) and turn west (left) onto Oronoco Street. Continue four and a half blocks until you come to Robert E. Lee's boyhood home (Figure 25). Enter the house and take the free tour. Note that the house across the street, which has a historical marker, was Lee's father's house. Continue west across Washington Street and note Lee's uncle's house on the other side of the street, which also has a historical marker, then continue for another block on Oronoco

Street (Figure 26), turning south (left) on Columbus Street (Figure 27.)

Continue south for three blocks until you come to Christ Church, George Washington's church (Figure 28). Go inside the church to see Washington's pew, linger in the book shop, exit the church, and go west (straight) on Cameron Street (Figure 29). Go six blocks and turn south (left) on West Street. Go five blocks until you come to a residential area that is closed to vehicular traffic. Continue straight through the townhouse development until you come to the Wilkes Street Cemetery Complex, a collection of nine small cemeteries, including the Douglass Freedmen's Cemetery. Spend a few minutes wandering among the historic Civil War-era graves (Figure 30).

Exit the cemetery complex on Wilkes Street and continue east (straight) toward Patrick Street (Figure 31). Cross Patrick Street, and continue straight on the pedestrian-only path. Begin your aggressive phase by stairstepping here. Turn north (left) on St. Asaph Street for one block, east (right) for one block on Wilkes Street, north (left) on Pitt Street for one block, east (right) on Wolfe Street for one block, north (left) on Royal Street for one block, east (right) on Duke Street for one block, north (left) on Fairfax Street for one block, and east (right) on Wales Street (Figures 32 and 33.)

Continue straight on Wales Street for two blocks to your operation act at the Old Dominion Boat Club (Figure 34), then continue to the waterfront to end your surveillance detection route.

Alexandria Virginia surveillance detection route.

Figure 19.

Figure 20.

Figure 21.

Figure 22.

Figure 23.

Figure 24.

Figure 25.

Figure 26.

Figure 27.

Figure 28.

Figure 29.

Figure 30.

Figure 31.

Figure 32.

Figure 33.

Figure 34.

About the Author

JOHN KIRIAKOU IS A FORMER CIA counterterrorism officer, former senior investigator for the Senate Foreign Relations Committee, and former counterterrorism consultant for ABC News. He was responsible for the capture in Pakistan in 2002 of Abu Zubaydah, then believed to be the third-ranking official in al-Qaeda. In 2007, Kiriakou blew the whistle on the CIA's torture program, telling ABC News that the CIA tortured prisoners, that torture was official US government policy, and that the policy had been approved by then-President George W. Bush. In 2012, Kiriakou was honored with the Joe A. Callaway Award for Civic Courage, an award given to individuals who "advance truth and justice despite the personal risk it creates," and by the inclusion of his portrait in artist Robert Shetterly's series, *Americans Who Tell the Truth*, which features notable truth-tellers throughout American history. Kiriakou won the PEN Center USA's prestigious First Amendment Award in 2015, the first Blueprint International Whistleblowing Prize for Bravery and Integrity in the Public Interest in 2016, and the Sam Adams Award for Integrity in Intelligence, also in 2016. A second portrait, by the noted Chinese artist Ai Weiwei, is in the permanent collection of the Smithsonian Institution.

Kiriakou is the author of multiple books on intelligence and the CIA. He lives with his family in Northern Virginia.